THE GREAT MISCALCULATION

THE GREAT MISCALCULATION

THE RACE TO SAVE NEW YORK CITY'S CITICORP TOWER

MICHAEL M. GREENBURG

WASHINGTON MEWS BOOKS

NEW YORK UNIVERSITY PRESS

New York

WASHINGTON MEWS BOOKS

An Imprint of

NEW YORK UNIVERSITY PRESS
New York
www.nyupress.org

Library of Congress Cataloging-in-Publication Data
Names: Greenburg, Michael M., author.
Title: The great miscalculation : the race to save New York City's Citicorp
tower / Michael M. Greenburg.
Description: New York : New York University Press, [2025] | Includes
bibliographical references and index.
Identifiers: LCCN 2024031651 (print) | LCCN 2024031652 (ebook) | ISBN
9781479829972 (hardback) | ISBN 9781479829996 (ebook)
Subjects: LCSH: Building failures--New York (State)--New
York--Prevention--History--20th century. | Real estate development--New
York (State)--New York--History--20th century. | 601 Lexington Avenue
(New York, N.Y.) | LeMessurier, William J.
Classification: LCC TH441 .G74 2025 (print) | LCC TH441 (ebook) | DDC
690.109747/10904--dc23/eng/20250304
LC record available at https://lccn.loc.gov/2024031651
LC ebook record available at https://lccn.loc.gov/2024031652

This book is printed on acid-free paper, and its binding materials are chosen
for strength and durability. We strive to use environmentally responsible
suppliers and materials to the greatest extent possible in publishing our
books.

The manufacturer's authorized representative in the EU for product safety
is Mare Nostrum Group B.V., Mauritskade 21D, 1091 GC Amsterdam, The
Netherlands. Email: gpsr@mare-nostrum.co.uk.

Manufactured in the United States of America

10 9 8 7 6 5 4 3 2

Also available as an ebook

William LeMessurier. *Photo courtesy of William Theon*

For Joyce

The potential danger would not have been discovered. . . . LeMessurier might have shrunk from the horror of disclosure and gambled that the building would somehow survive. If either of these possibilities had occurred, it is as certain as anything in the state of the art of engineering and mathematics that Citicorp Center had a 100% probability of total collapse by the end of the century. . . . When collapse occurred, it would have come suddenly, without warning, and would have killed thousands of people.
—William LeMessurier, "Project SERENE," November 20, 1978, 33–34, Papers of William LeMessurier

Little pig, little pig, let me come in.
No, not by the hair on my chinny chin chin.
Then I'll huff, and I'll puff, and I'll blow your house in.
—Jacobs, *English Fairy Tales*, 69

CONTENTS

PROLOGUE

"ANCHOR OF SERENITY"

It was the eighth time in the congregation's 111-year history that the parishioners of St. Peter's Lutheran Church solemnly packed their sacramental vessels and wandered from their home in the borough of Manhattan. The latest migration, in the winter of 1973, was a nod to unrelenting progress.

Closing the ancient wooden doors of the now crumbling Victorian structure at Fifty-Fourth Street and Lexington Avenue for the final time, a throng of over five hundred congregants trailing celebratory streamers and balloons marched to Park Avenue and began the ten-block journey north to Central Presbyterian Church, their latest temporary place of worship. Behind them, the St. Peter's bell that had tolled in greeting so many times in the past now bade them somber farewell. "It is proper that we should take to the streets again," Pastor Ralph Peterson told the assemblage. "We are a church on the move."[1]

Central Presbyterian's congregation, led by its pastor, Rev. Dr. Robert A. Edgar, had graciously opened its doors to the St. Peter's flock. On the steps of the sprawling Victorian building commissioned in 1921 and funded almost exclusively by John D. Rockefeller Jr., a jazz ensemble gaily playing "When the Saints Go Marching In" welcomed the parade of newcomers to their provisional home.[2]

While the stay at Central Presbyterian was anticipated to be short-lived—just a few years while a new church could be built—not everyone at St. Peter's was pleased. "We don't like it at all—this is the second time I'm being thrown out," said eighty-two-year-old Bertha Pfeiffer, smiling weakly as she struggled to catch her breath after the lengthy walk up Park Avenue. Mildred Westermann, who dotingly transported the St. Peter's fair-linen altar cloth to its new location, said, "I just didn't look back at that beautiful church," as she gulped back tears.[3]

She was there when the "new" St. Peter's opened its doors in 1905.

◢

St. Peter's Lutheran Church had a long and rich connection to its Midtown East neighborhood. The grand, neo-Gothic cathedral endured as a symbol of Christian faith and tradition for much of the twentieth century. A story of growth and adaptation in Manhattan, the church became part of the religious and cultural landscape of the city. "St. Peter's stands at the center of this multi-layered intersection, intent, as always, on joining and shaping this ongoing conversation," observed its current senior pastor.[4]

The Deutsche Evangelische Lutherische Sanct Petri-Kirche, as the congregation was formally christened in 1862, established its first meager house of worship in a small garret above a feed and grocery store in Manhattan, not far from what would later become its permanent home. "Both rooted and nomadic," the church was founded by a small band of German immigrants searching for religious expression and escape from declining political and economic conditions in Europe.[5] The group was part of a mid-nineteenth-century explosion of German immigration to the US and to New York in particular. By 1860, more than two hundred thousand Germans inhabited New York and accounted for about one-quarter of the city's total population.[6] Religiously diverse, most early émigrés were Calvinists, but later Catholics and Lutherans joined them throughout the area.

Sanct Petri-Kirche grew rapidly during its first ten years and required several moves to new and larger facilities: one above a butcher shop and others in space abandoned by various parishes also on the move. Sometimes the fledgling sect simply moved in with other churches and temporarily combined their homes as they waited for more suitable accommodations to become available.

By the 1870s, Sanct Petri-Kirche had purchased a larger Gothic Revival building in Midtown Manhattan, where, gradually, its congregation assimilated and a liturgy in English was added. Embracing a growing and more culturally mixed patronage, the church sold its property in 1903 to the New York Central Railroad for use in the construction of Grand Central Terminal and commissioned its seventh and final home: a new, grandiose building at Fifty-Fourth Street and Lexington Avenue that even Stanford White, the famous architect of the Gilded Age, would have envied.

"The project was the most opulent and complex the parish had ever undertaken," wrote one architectural historian.[7] Designed by John C. Michel, a member of the St. Peter's Young Men's Association, the classically Gothic church featured a towering spire at the corner, beneath which gabled limestone wings wreathed with stained glass stretched along each city street. Several turrets stood guard on the jagged roofline like medieval centurions, and interior groin vaults reminiscent of Charlemagne's Palatine Chapel in Aachen, Germany, expressed artistry in a sea of stone. The building's main sanctuary and balcony housed 650 worshipers, who gazed on a chancel gilded with carved wooden sculptures and, beyond, an immense arched mural magnificently depicting the Sermon on the Mount.

In this resplendent place of worship, St. Peter's remained for nearly seventy years.

◢

The "progress" that St. Peter's ultimately yielded to in surrendering its longtime home arrived in the form of two young and determined real estate brokers named Don Schnabel and Charles McArthur.

One Saturday morning in the early fall of 1968, the two "well-tailored" brokers stepped out of a yellow cab on Lexington Avenue in front of St. Peter's Church and, rather conspicuously, began surveying the shops and businesses along what would later be called "the most expensive block in New York's history."[8]

The weather had been unseasonably warm, and Schnabel and McArthur, dressed in their plaid wool suits, began to perspire despite their deliberative pace. They held small notebooks, and as they strolled past Howard Johnson's, Save-Mor Drugs, Lexington Sandwich Shoppe, Carroll's Pub, and Anthony's Restaurant, they surreptitiously scribbled addresses, business names, and frontage lengths. They paused at the Medical Chambers on Fifty-Fourth Street, "the handsome but dowdy doctors' building in the middle of the block," and they whispered to each other presumed estimates of value and square footage.[9] Making their way past several attractive brownstones, they reached the city corner where they had begun, stopped, and, like hunters stalking elusive prey, gazed pensively upward at the ancient and crumbling steeples of St. Peter's Church.

Schnabel and McArthur had done their research. They worked for the very prestigious brokerage firm of Julien J. Studley, Inc., and they had an abundance of contacts and resources at their fingertips. They learned that of the thirty-one parcels that made up this "intricately small-scaled, richly diverse block," as one observer described it, most had been held by the same owners for years.[10] There had been no attempt, as far as they could tell, to "assemble" the properties on behalf of one client for a single combined—and massive—project. This block appeared to be the perfect target for just such a strategy.[11]

But the scheme had to be done quietly. In similar endeavors, despite the most clandestine development plan and disguised investor, word almost always leaked. When the extent of the plan was revealed, prices rose like a fever. So the two brokers went about their business unobtrusively, patiently, making contacts and probing building owners for opportunities, never divulging who their client could be.

◢

The early decades of the twentieth century seemed filled with promise for St. Peter's. With its new prominent home and growing congregation, the church survived and flourished through two world wars and the Great Depression. Predominantly English speaking and aware of the pervasive anti-German bias during and beyond the war years, the church formally changed its name in 1925 to "Saint Peter's Lutheran Church of Manhattan."

By midcentury, the city had been transformed. In an explosion of Midtown construction, granite skyscrapers replaced small businesses and quaint brownstone residences. As the city prospered, however, congregants fled into New York's outer boroughs and suburban communities, and participation in downtown churches began to decline. St. Peter's was no different. Once a thousand strong, its membership dwindled to less than three hundred by the mid-1960s. "All in all," observed the author Peter Hellman, "St. Peter's was like so many other city churches. It was dying."[12]

When Dr. Ralph E. Peterson was called to serve as the seventh senior pastor of St. Peter's in 1966, the church had been barricaded behind a wrought-iron gate, open only on Sunday mornings for an ever-shrinking congregation. The sagacious forty-two-year-old Minnesotan, however, brought an advanced degree from Harvard and a brash vision of growth and identity to his new home. Through his work at the National Council of Churches in the early 1960s, Peterson embraced the notion that religion must be a part of contemporary culture, including the arts, music—and jazz.

Believing that art should "speak in theological judgement," the young minister insisted that the Reverend John Garcia Gensel, a devotee of theological celebration through music, join him at St. Peter's to create a "jazz ministry."[13] The idea was initially met with skepticism and downright indignation by some parishioners; but the introduction of music gradually brought renewed interest in the church, and the uncertainty melted away. Soon, the literal iron gate insulating St. Peter's was unlocked, and the congregation opened its doors to the people of Manhattan day and night. Gensel recognized that many musicians who worked late on Saturday nights would welcome a later religious service on Sunday evenings. They became known as "Jazz Vespers."

By 1968, Peterson and Gensel's concept had taken wing. Jazz luminaries such as Duke Ellington, John Coltrane, and Dizzy Gillespie performed regularly at what would soon be called "The First Church of Jazz" and then simply "The Jazz Church." Ellington would satirically dub Reverend Gensel "The Shepherd of the Night Flock."[14]

"Jazz has soul," said Peterson.[15] His new ministry would provide "a God-given opportunity to create in midtown Manhattan a vibrant and dynamic locale in which to reopen the dialog between Church and the arts."[16] It was, according to Peterson's favorite expression, an invitation to be "more human in skyscrapers."[17]

Even as the pastor was culturally enriching his church, the economy of New York, and of the country as a whole, was declining. Vietnam and President Johnson's War on Poverty had drained government resources, and inflation began to erode the prosperity of the 1950s and 1960s. Manhattan, an economy unto itself, was in a state of transition. Many urban real estate developers had receded into a conservative "wait and see" posture, but some forward-thinking speculators contemplated expansion. Peterson remained optimistic about the area surrounding St. Peter's, but despite a renewed interest in his ministry, demographics did not favor enduring growth for the church. The early-century church building was still crumbling, and the congregation gave serious thought to selling the land and relocating to the new United Nations district by the East River or elsewhere.

But Peterson understood the value of St. Peter's present location, and he knew that the land may be prized by commercial developers. His church, he shrewdly concluded, could one day be leveraged into valuable opportunity. According to Peterson, it was "a funky in-between neighborhood" full of restaurants, bars, shops, and residences.[18] Bloomingdale's was just four blocks north, and the Waldorf Astoria was virtually next door. Fifty-Fourth and Lexington had been the home of the church for over half a century, and it was there that Peterson and his flock would stay. "Christians have always been in the heart of the city," he wrote in a church mission statement. "The role of the church is not to retreat into the catacombs. . . . We have a destiny to be fulfilled here."[19]

Yet, at the end of every day, as Peterson descended the steps of his church, a gentlemen's lounge directly across the street called "The Bottomless Pit" stared him in the face.

It was a fitting metaphor for the perils that lay ahead.

◢

Don Schnabel had already approached a very interested candidate to develop the property, and he obtained the necessary authorization to proceed. He contacted several key owners in the neighborhood and began to gauge interest and willingness to sell. Slowly, skillfully, he developed relationships and brokered negotiations. He formed a holding company—Lexman Realty Corp.—owned by one undisclosed shareholder—to purchase and own the real estate, and gradually Lexman began acquiring individual parcels on behalf of its client.

The process would prove arduous and expensive. Later, as word emerged that the principal owner of Lexman might be a high-powered business concern and that a massive construction plan was envisaged, negotiations predictably stiffened. Along with the real estate came existing leases that had to be purchased, and soon emboldened owners and tenants dug in. *Town and Country* wrote, "One old woman yielded her lease when, along with money, the [suitor] threw in an airline ticket to California and agreed to drive her to the airport in a limousine."[20] The owner of a lounge with a long-term lease on the block had agreed to a $200,000 buyout with Lexman. Upon the owner's learning of the plans for the area, the asking price rose to $400,000. And, after lengthy negotiations, the forty doctors who owned the Medical Chambers building on Fifty-Fourth Street finally exchanged their interest in the property for $7 million of the acquiring company's stock. The total acquisition cost of the block would ultimately exceed $40 million.[21]

Early on, the St. Peter's property had become an ongoing and critical piece of the acquisition plan. In 1968, a wealthy parishioner of St. Peter's bequeathed to her church three adjacent, contiguous parcels of land. Ralph Peterson quickly recognized the significance of the gift as providing, in his words, a "major stranglehold on the block": "We had the corner," and no one could develop the area without including the church in the arrangement.[22]

In the fall of 1969, as the value of the church's "asset" became fully apparent to its pastor, Schnabel made contact with Ralph Peterson, and the two men arranged lunch at a local pub. After an exchange of pleasantries, Schnabel came to the point. "How much would it take to get you to move elsewhere?" "Why move elsewhere?" asked Peterson with a sardonic grin. "This is the best place to be." He explained that St. Peter's had been born in the city and was not about to be driven out by commercial development. He leaned forward and, with good-natured sternness, added, "So if you're going to do anything, you better damn well include us in it."[23]

Peterson well understood the enviable position the church found itself in, but he also knew the limits of his own abilities. To assist in the complex negotiations, he hired a well-respected Manhattan realtor named John R. White, who would later acquire fame for his nimble structuring of the Pan Am Building sale in a manner that ingeniously—and legally—avoided the payment of a $125,000 transfer tax to the city of New York.[24] Peterson made it perfectly clear to the agent that the church had no intention of moving from its present location, but he also confessed that its home, the old Gothic building, was in terrible disrepair. Perhaps, suggested Peterson, an accommodation could be made to accomplish the goals of all parties.

In the coming weeks, a very heated negotiation between the real estate agents took place. "There was a great struggle," the pastor later recalled.[25] After Schnabel had consulted with his client, he offered to place the church on the top floors of whatever building would be constructed. Peterson rejected that. He wanted to be where the people were, not in the clouds. Next, Schnabel offered that the church be placed in a storefront window at the base of the building. Peterson also rejected that. "We wanted our own separate identity, quite visible, quite unlike the tower," he later said.[26]

Meanwhile, at the time, neither Peterson nor White had any idea whom they were actually negotiating with. Schnabel, in his effort to maintain total secrecy, had never disclosed his client's identity or the purpose for which it wanted the church land and the remaining parcels on the block.

That changed in early February 1970.

⁂

One morning, as the brokers wrangled over terms and conditions, Schnabel informed Peterson that his client would like to personally meet him. Later that evening, an emerald-green Cadillac limousine slowly emerged from the darkness and stopped in front of Peterson's home. In the back sat a handsome, middle-aged man dressed in a gray Brioni business suit. The chauffer doffed his hat, and as Peterson slid in, the gentleman smiled disarmingly and handed him an embossed business card. It read, "Henry Muller, Senior Vice-President, First National City Bank."[27]

Though rumors had swirled, Peterson had no real idea that the party who had been quietly courting the church was already its neighbor. At the time, Citibank, as it had come to be known, and its parent company, First National City Corporation (or Citicorp), occupied the old office building at 399 Park Avenue, about a block away from St. Peter's.[28] Built on land once owned by John Jacob Astor, the timeworn edifice that was home to the organization for much of the decade no longer conjured the eminence of the economic powerhouse that Citicorp had become.

Originally chartered in 1812 under the name City Bank of New York, the company had, through the years, emerged as one of the nation's largest and most powerful financial institutions. Described by *Fortune* magazine in 1965 as the "wave of the future" for US banking, Citibank had grown to 150 local branches and 163 international branches, positioning itself as the leading bank in the New York metropolitan area and the second largest financial institution in the country.[29] Yet, by the late 1960s, the company, though on par with its most prominent corporate competitors, lacked a signature command center emblematic of its muscular reputation. When Schnabel and McArthur first recognized the economic potential of the city block as an assembled whole, they approached the bank and quickly found a willing partner.

As Henry Muller arrived at Peterson's residence in his stretch limousine on that winter night in 1970, he was one of merely five or six people in a company of forty thousand employees who had even the slightest inkling that Citicorp was on the move.

Peterson was still processing the implications of the bank's involvement when the men arrived at Delmonico's in the financial district for a private dinner. Muller confessed that it was he who was the enigmatic principal behind Lexman Realty Corp. and that the bank had, in fact, partnered with the two brokers from Julien J. Studley to consolidate the block for development. Their purpose, he explained, was to build a preeminent skyscraper on the property befitting of the Citicorp name.

Muller was educated in New York as a real estate specialist and served in the South Pacific with the US Army during World War II. He began his civilian career as an engineer for Prudential Insurance Company and worked for eight years as deputy director of buildings and grounds at Harvard University. In 1964, he accepted a position with Citibank as a senior vice president and ultimately attained the chairmanship of the bank's real estate division in New York City. He was engaging and affable—but, more importantly, he was a Lutheran. The pastor immediately took a liking to him.

Peterson impressed on the executive that if an agreement was to be reached, it must respect the structural autonomy of the church in its current location. The old Gothic building would have to be demolished, and in its place, a new, visible, and creatively distinct church must be constructed. Muller agreed that, perhaps, the bank could work around the space constraints presented by Peterson's demands, but it would take ingenuity to pull it off. Though the details were left to future meetings, the two men recognized that with "vision and fiscal reality" an agreement might be in reach.[30]

In the coming days, Peterson's agent, John White, went to work. Fueled by the basic understanding reached by his client and Henry Muller, he considered every possible legal and architectural scenario to accommodate both Citicorp's skyscraper and St. Peter's new church in the same confined space. Finally, in a moment of revelation, White realized that if the church were to sell its land to Citicorp, which would then convert the entire block into one large condominium, the bank could sell the corner back to the church as a condominium unit, thereby retaining an agreed portion of the air space above for its skyscraper.

The idea took wing, and several days later, Ralph Peterson signed a letter of intent to sell the land on which St. Peter's had sat for generations to Citicorp for $9 million. Under the agreement, the bank would demolish the decaying present structure and on the same parcel construct a new free-standing church at a cost of $5 million, leaving St. Peter's with a $4 million profit. "Since St. Peter's real estate clout was equal to its moral clout," wrote the *New York Times*, "a little arm twisting might have been involved."[31] The final agreement was overwhelmingly approved by the voting members of the congregation.

With this arrangement in place, Peterson envisioned a "'teaching parish' combining seminary and an area for . . . students to work in; an international chapel and information center for visitors; a meeting place where the old and the lonely can come; a day-care center for working mothers; theaters and art galleries. . . . The main thing," he said, "is to create a living room, . . . a place where people can come and be human for a while."[32]

In keeping with Peterson's vision of an architecturally independent building, physically separate from the skyscraper, it was decided that the air space above approximately two-thirds of the footprint of the new church would remain open and that no element of the bank tower would encroach on or interfere with the church building below. Unambiguous and seemingly innocuous, these requirements would set into motion an engineering predicament with potentially catastrophic implications.

In 1971, Pastor Ralph Peterson described the soon-to-be-constructed St. Peter's church as "an anchor of serenity in—but not a withdrawal from—the sea of unpredictable turmoil around us."[33] The statement would turn to prophecy.

1

A SKYSCRAPER ON STILTS

Cambridge, Massachusetts, was, as always, a college town. Through its cobbled streets and fabled courtyards strolled Roosevelts and Rockefellers, Kennedys and Kissingers, Ginsbergs and Galbraiths. Harvard, the intellectual heart of the city, pulsed with innovation, revolution, and leadership. Founded as a religious training ground for young men of the Christian ministry, the college would evolve into the world's leading educational institution, transforming Cambridge into a celebrated hub of culture and academia.

Along the city perimeter, and at the doorstep of Harvard, flowed the river named for the English king who claimed it. The Charles separated Cambridge from Boston to the east and, with its tidal shifts, provided the early settlement protection from pirates and invaders. To Harvard, the river would become little more than a polluted and fetid mudflat. "The Charles was Harvard's humble backdoor, functioning as a kind of loading dock for food, coal, and lumber," wrote an editor of the *Harvard Gazette*. "Sewage was rampant—a legacy of pollution that eventually closed the river to swimmers in 1955."[1]

But Harvard would learn to embrace its neglected waterway. As early as the 1880s, a gifted landscape architect named Charles Eliot, protégé of the famed designer Frederick Law Olmsted, initiated public debate about the plight of urban environments and the need for natural milieus in US cities. One historian wrote of the result, "After 40 years of debate, the Charles was dammed in 1910, shutting out the sea's salt water and eliminating tides. Overnight, the vast porous basin was transformed into 'a big, long lake. . . .' The river became a shimmering mirror in which Cambridge and Boston could see themselves anew."[2]

It would take the better part of a century before any serious measures to cleanse the river of sewage and waste would take hold, but the Charles would no longer be the blight of metropolitan Massachusetts.

In the late nineteenth century, as the US flexed its industrial might, urban powerhouses arose, as did artistic expression in building construction and design. Following the lead of an emerging discipline, Harvard began teaching courses in architecture in 1874. By 1936, the Graduate School of Design (GSD) was formally inaugurated, uniting the three disciplines of architecture, urban planning, and landscape design in one advanced degree program. Among the early graduates of the Harvard GSD was a tall, athletic man—a nationally ranked collegiate track star and Olympic hopeful—by the name of Hugh Stubbins Jr.

Born in Birmingham, Alabama, in 1912, Stubbins displayed an early penchant for art and design as well as athletics. He attended the Georgia Institute of Technology and earned a master's degree in architecture from the Harvard GSD.

In 1939, Walter Gropius, a pioneer of the modernist movement in US architecture and a professor at Harvard, invited Stubbins to become his assistant and to join the graduate school of design faculty. Stubbins moved to nearby Lexington and while teaching set up his first architectural practice, primarily focused on modern single-family homes in New England—"sleek compositions of solid planes and sheets of glass, along with regional materials."[3] Upon Gropius's departure, Stubbins became the department chair at Harvard GSD, where he remained until 1954. Later devoting himself exclusively to his growing private practice in Cambridge, he would work, and later live, in the city by the Charles for the remainder of his career.

Stubbins's first major architectural project came in 1957 with his design of the Congress Hall in West Berlin—the US contribution to the International Building Exhibition. The technically innovative building quickly earned the moniker "the pregnant oyster" for its arching and elongated contours.[4] Stubbins later designed a variety of academic buildings for Harvard, the Massachusetts Institute of Technology (MIT), Princeton, and the University of Virginia, and in 1967, his firm, Hugh Stubbins and Associates, was named the American Institute of Architects firm of the year.

Throughout Stubbins's six-decade career, he would be named as the lead architect for hundreds of domestic and international projects, including Veterans Stadium in Philadelphia, the Yokohama Landmark Tower in Japan, and the Ronald Reagan Presidential Library in Simi Valley, California. "His architectural sympathies were modern in every way," wrote one architectural historian. "Stubbins was focused on function, on problem solving, on planning. Architecture had to 'work,' responding to the needs of human beings."[5]

In 1958, Harvard University commissioned Stubbins to design the new Loeb Drama Center on Brattle Street in Cambridge. The facility would become home to the American Repertory Theater and the Harvard-Radcliffe Dramatic Club, but during the planning and construction phase of the project, it played the more pedestrian, yet fateful, role of introducing the architect to the deputy director of buildings and grounds at Harvard: Henry Muller. As the men worked together throughout the project, Muller would come to admire Stubbins for his meticulous nature and technical expertise, and through the years, the men would maintain a genial acquaintance.

A decade later, as Muller finalized his deal on behalf of Citicorp with Ralph Peterson and St. Peter's Church for that essential corner of the assembled block on Fifty-Fourth and Lexington, the work began to envisage the bank's signature headquarters. "It was an act of optimism and renewal of faith in a climate of economic uncertainty," wrote Stubbins's biographer of the bank's conviction.[6] While many businesses were at the time fleeing the blight, crime, and financial decay of the inner city, Citicorp was expressing its confidence in and commitment to Manhattan. The company's new home needed to reflect the boldness of that commitment and of the bank's forward-thinking philosophy as a whole.

Muller had consulted with the Citicorp board of directors and, no doubt, with its chairman, Walter Wriston, regarding the choice of architect for the project, and he avidly promoted Stubbins. His firm had recently undertaken several public and private high-rise residential projects, as well as the State Street Bank Building in the Financial District of Boston and the Federal Reserve Bank of Boston. His designs had become increasingly sophisticated, and his reputation as a prominent architect was emerging.

In the summer of 1970, Muller began preliminary discussions with his old acquaintance from Harvard for the design of the new Citicorp world headquarters. At the outset, Muller conveyed one unbendable condition: the plans had to accommodate—and could not compromise—the land space currently occupied by St. Peter's Church, as required by the existing contract with Pastor Peterson.

It was clear to Stubbins from his initial meetings with the Citicorp managerial team that the company was looking for more than just another concrete monolith on the Manhattan skyline. The bank's corporate responsibility to the community seemed to imbue their conversations. Perhaps, urged the team, Stubbins might conceive not just an office tower but a cultural contribution to the city itself.

◢

The New York City zoning bylaw of 1961, in fact, incentivized such a mind-set. The purpose of the act was to encourage developers to create and enhance public-private open spaces on the land surrounding their buildings, in exchange for expanded floor areas within their new structures.[7] Literal compliance, however, had not historically provided the desired result. Most developers fulfilled the letter of the law to gain its benefits but ignored the spirit. "[They] instructed their architects to design 'keep off' space without seating, trees, or amenities of any kind. Some of this so-called public space [was] hidden behind locked gates and unmaintained, and almost none of it was designed in such a way that the public [felt] free to use it," wrote one observer.[8] These "open space" plazas were "bleak, forlorn places. . . . Some [were] forbidding and downright hostile," wrote another.[9]

Even as New York mayor John Lindsay's "Urban Design Group" sought to clarify and strengthen the 1961 zoning resolution (which it did in 1975), Citicorp had already gotten the message. Though the immediate economic outlook of the neighborhood was uncertain, the bank nonetheless understood that 40,000 people lived and more than 160,000 people worked within an eight-minute stroll of its city block.[10] The area had historically been a flourishing shopping district, and the executive team was enthusiastically open to the idea of creating a lively multi-tiered combination of open space and retail as a companion to its office tower on the site.

Stubbins immediately signed on to the concept. He understood the need for a design that combined aesthetic impact with cultural appeal. The Citicorp Center, he believed, must, above all, represent corporate citizenship. In a September 1970 "pitch" letter to Henry Muller regarding what he called the "office building/ church project," Stubbins wrote, "The new, slick, slab buildings that march up the avenues of New York and other U.S. cities are symbolic expressions of the Machine. They are anonymous—cool and inhumane. We need to develop a new generation of Office Buildings, rooted in planning for the community and expressive of the human individuals who use and control them."[11] Later, he elaborated his vision for the project to a reporter:

> By revitalizing urban development with an emphasis on people, we could produce a more enjoyable place in which to live and work. . . .

With the church as catalyst and the bank as supporter, we can design a new kind of place which all kinds of people will want to enter and become part of. While the church must have its own identity, I like to think how it could be enhanced and magnified if we combine it with a new kind of office building.[12]

Despite Stubbins's outward enthusiasm, he had a nagging concern about his lack of experience in the Manhattan market, and he worried that the bank may as well. "One might wonder why we as Cambridge architects are interested in this New York City project," he wrote to Muller. "Very simply," he continued, "we believe we have the talent, the experience and desire to make a real contribution to the most important center of culture and communication in the world. We believe that today architecture must be a form of communication. We are anxious to work for and with people who have the courage to be first—to lead and not to follow."[13]

The truth was that Stubbins had never designed a high-rise structure in New York City before. As competent an architect as he was, he was regarded as rather traditional in his thinking—"one of the old line conservative architects," according to one commentator—and this project, he knew, would require innovation and imagination.[14] It was an unfamiliar regulatory landscape, and the Citicorp Tower was the most ambitious and complicated commission he had ever embarked on.

And there was the church.

William LeMessurier often lunched at the same Greek restaurant in Cambridge. The Acropolis, on Massachusetts Avenue, specialized in classic Mediterranean cuisine and was just steps from the structural engineering firm that bore his name.

Shortly after Stubbins began conversations with Citicorp, he called on LeMessurier and asked if he might join him for lunch. The two men had known each other since the early 1950s and had worked on many projects together through the years, most recently the two high-rise bank buildings in Boston. Stubbins became one of the engineer's "first and longest-lasting architectural clients," said William Thoen, LeMessurier's former partner. "As soon as Hugh and Bill met, there was a chemistry between them. Both were looking for excellence in their work. . . . LeMessurier became Hugh's only structural engineer. Stubbins was sort of destiny's [lot], and as his reputation grew, so did ours."[15] LeMessurier himself would describe Stubbins as "a formal man, a southern gentleman" and their association as "a life long satisfaction."[16] With the Citicorp project, however, Stubbins had a problem—and with Bill LeMessurier, he was hopeful he would have the solution.

Originally from Pontiac, Michigan, William LeMessurier was fourteen years younger than Stubbins but, like the architect, displayed an early proficiency in mathematics and the arts.[17] He earned a bachelor's degree in math from Harvard and, as an admirer of Stubbins, decided to pursue graduate work in his footsteps. "I first knew of Hugh's work when I was a youth interested in architecture, and his presence at the Harvard Graduate School of Design was one of the attractions which led me to study there," he wrote in 1977.[18]

LeMessurier, realizing that he was more interested in the technical and mathematical aspects of the field, soon left Harvard to study structural engineering at MIT. As a student, he worked part-time for the noted Boston structural engineer Albert Goldberg, who hired him as a partner in the firm after graduation in 1953. Eight years later, LeMessurier formed a new and independent partnership in Cambridge—LeMessurier Associates—which he maintained in various forms for the duration of his engineering career.

Of LeMessurier's talent, a colleague noted, "Bill pioneered the use of innovative structural systems that efficiently resisted gravity, wind, and earthquake loads while respecting the aesthetic concerns of his architectural clients."[19] He served on the Boston Building Code Committee and was instrumental in drafting amendments to the state and local building codes.

LeMessurier would become a highly respected and published professor at both MIT's Department of Civil Engineering and the Harvard GSD, and he would earn a variety of professional honors throughout his career, including the Allied Professions Medal from the American Institute of Architects, appointment to the American Institute of Steel Construction Committee on Specifications, and election to the National Academy of Engineering.[20] "Bill loved teaching as much as engineering, and was always at his best with an audience," recalled Thoen. "He was extremely intelligent, insightful and highly articulate, and if you got into a verbal argument with him you would surely lose, usually in the first round. He thought very carefully about whatever he said and was precise in his use of language. I think that is what made him such a good leader, lecturer and teacher."[21]

LeMessurier excelled artistically as well scientifically. He was an accomplished pianist and was, as a college student, a member of Harvard's Glee Club. Later in his adopted hometown of Wellesley Hills, he delighted his family with sing-alongs and showcased his dramatic flair by performing at the local playhouse. He was incessantly meticulous and endearingly eccentric, but despite his numerous technical and creative accomplishments, he could not hide a nagging inner fear of failure that would plague him for most of his life.

Seated for lunch at the Acropolis, Stubbins chewed on a cigar and peered intently at LeMessurier. He described his long-standing relationship with Henry Muller at Harvard and how the two had recently been talking about the Citicorp project. It was a venture that he was eager to be a part of, but he needed LeMessurier's expertise. Stubbins suggested that partnering together on the project, as they had many times in the past, could result in a lucrative opportunity and a stepping-stone for both of them into the New York building-design market.

The unique engineering challenge—and the overriding concern that would drive the design of this complex—Stubbins continued, was St. Peter's Church. The church occupied a critical corner of the building site, he explained, and though it would be rebuilt as part of its agreement with the bank, it would remain at precisely the same location. To further complicate the situation, the church building had to be free-standing, intruded by no portion of the tower, and with the air space above at least 63 percent of the structure left open.

LeMessurier's brow furrowed. In a typical square or rectangular high-rise construction scenario, the building's bottom support columns would logically be positioned at the four corners for optimum strength and even distribution of wind and gravity load. The problem with the Citicorp project, as he understood it, was that given the required location of St. Peter's Church, one column of the skyscraper, traditionally placed at its corner, would run directly through the church sanctuary.

Stubbins had appreciated this problem and had initially considered notching or cutting the column or even the building itself to make room for the church.[22] Failing that, he asked the engineer whether one corner of the building could be extended or "cantilevered" beyond the support columns, leaving sufficient space below for the church. LeMessurier nodded and reminded Stubbins that several years earlier they had designed something very similar into the State Street Bank tower, where cantilevers were extended seventeen feet beyond the edges of the building at each floor to preserve the view. As the men talked, they realized that if they could cantilever one corner of the Citicorp Building, why not two or three—or even all four.

While they waited for their food to arrive, LeMessurier took a pen from his breast pocket and began sketching on a paper napkin at the table. His drawing depicted an intricate structural bracing system with ground-level support columns intriguingly positioned at the *middle* of each face of the building, rather than at the corners, thus allowing a "plaza" below for the church as well as other uses. "There's a point of near ecstasy when you make a discovery like that and you draw it on a napkin and that's that," he later said.[23]

In essence, he had envisioned a skyscraper on stilts.

2

"SCARY EXCITEMENT"

By the early years of the "Me Decade," as Tom Wolfe musingly dubbed the 1970s, Manhattan had become a city on the brink of chaos. Stained by economic malaise, urban decline, and racial injustice, the once vibrant empire reeled with turmoil and despair.[1]

Budget deficits and declining industry decimated city services, while soaring unemployment gnawed at the fabric of New York's working-class neighborhoods. Forgotten and neglected quarters of the South Bronx, Upper Manhattan, and even Brooklyn quite literally burned, leaving in their ashes the malaise of bankruptcy and hopelessness.[2] Poverty, drugs, and desperation produced a breeding ground of lawlessness. New York's graffiti-shrouded subway trains—the artist's canvas of a dispossessed youth—shuttled frightened citizens past broken glass, barred windows, and abandoned buildings occupied only by drug dealers and rats. Throughout the decade, over 820,000 middle-class New York families fled the city for the refuge of suburbia.[3] As Jonathan Mahler wrote, "The clinical term for it, fiscal crisis, didn't approach the raw reality. Spiritual crisis was more like it."[4]

Yet, even in the midst of this "spiritual crisis," New York retained its sanguine and enduring essence. Ethnic diversity accentuated and strengthened the neighborhoods of New York in the 1970s. Artists from every realm, including literature, music, theater, and portraiture, blossomed in shops and playhouses from SoHo to the East Village. Dance clubs and disco bars energized the city's nightlife, uniting disparate communities with the vibrancy of music and social life. With the restoration of such legendary structures as Grand Central Terminal and the South Street Seaport, awareness of New York's architectural history was reborn. The Landmarks Preservation Commission, founded in the mid-1960s under the leadership of Mayor Robert Wagner, marshaled an attitude of renewal that would carry through the ensuing decades.

That Citicorp—and more fundamentally its intrepid chairman, Walter Wriston—chose to remain in the vicinity of Midtown Manhattan while others fled and to invest while others divested was described by one historian as "a noteworthy, if not astounding event."[5]

With acres of vacant office space blighting the city in the early 1970s, Citicorp was, to say the least, faced with an uncertain real estate market. The newly constructed Twin Towers complex of the World Trade Center, conceived as a government revitalization project for Lower Manhattan and financed by the New York and New Jersey Port Authority, was, at the time, in direct competition with the private sector for tenants and would not even be fully occupied until near the end of the decade. Citicorp chose a long-term view of its investment. Despite the turbulent and uncertain economic environment of the day, the company's leadership team believed that conditions would ultimately improve and that New York City would endure as a major hub of commerce and banking. Construction costs during the economic downturn, it was thought, were manageable and labor plentiful. The company aspired to the rejuvenation of the city skyline and sought to position itself as a catalyst for growth in Manhattan.

The competitive advantage of the prestigious location was not lost on the management team, however. The prominence of a resplendent skyscraper in the heart of the city would send a bold and symbolic message of corporate strength and stability and would solidify the company's position as a worldwide leader in the financial sector. In announcing the estimated $128 million project, Wriston said, "The new New York may not emerge tomorrow, but there is no question that is on the horizon. Citicorp Center is perhaps a sign of what it can be like and we think that is a very positive sign and expression of our very strong faith in New York and New Yorkers."[6]

From the start, the public face of the project was unmistakably planned to be community focused. The bank had stressed its desire to enhance the neighborhood without undue harm or inconvenience to its inhabitants. Wriston focused in his public statements not on the bank itself but on its social contribution to the community. "We believe New York is very much here to stay and Citicorp Center is our way of saying, 'So are we!'" he said. "It is designed to be a living, positive part of the neighborhood it stands in, twenty-four hours a day and for decades to come. We believe it is a giant step towards the goal of erasing and replacing what critics have so long described as 'the impersonal architecture of the city.' A visitor may occasionally get lost in Citicorp Center, but none will feel alone and isolated."[7]

Hugh Stubbins buttressed Wriston's sentiment. "We had a very enlightened client in Citicorp," he later said. "They wanted to change the image of corporate construction in New York and at the same time inject a vitality into the neighborhood that is too often lost in the congestion of high-rise structures."[8]

Whether believed or not by the residents and business owners of Midtown Manhattan, the bank's commitment to the social and cultural fabric of the city was going to be the resounding anthem of the project, echoing from inception to completion.

◢

By the fall of 1970, Hugh Stubbins had begun informal meetings with Henry Muller and the president of Citicorp, William Spencer, who was the direct liaison to Chairman Wriston. The bank had still not given Stubbins any formal commitment on the project, and he continued to coax Muller for a more concrete arrangement. "We have the talent and capabilities to make some imaginative and realistic suggestions for the development of the area, and I would hope that you could authorize us to spend 8 to 10 weeks to develop several alternative concepts," he wrote to Muller on November 2.[9]

Muller soon gave Stubbins the go-ahead, and in February 1971, after extensive research and careful consideration of the locale and the bank's primary objectives, Stubbins submitted an initial development and concept design study of the property. The bank had been considering a variety of potential site configurations, several of which also contemplated a new "Medical Chambers" building for the neighboring doctor's cooperative. Adding to the complexity was the ongoing land-assemblage effort on the block. Final negotiations with several key owners—including the doctors—were still unconcluded, and uncertainty remained as to the feasibility of the main tower's column placement and the structural base formulation. Over the next year, Stubbins conferenced tirelessly with the Citicorp team and conducted developmental studies for each of the site options. In February 1972, he provided the bank with a set of seven general alternative design schemes for four possible site configurations.

Throughout the design process, Stubbins relied on William LeMessurier and the engineer's young associate, Robert McNamara, for assistance and advice on the structural elements and costs of the project. A recent graduate from UC Berkley, McNamara brought a modern and youthful vitality to LeMessurier's office and would prove instrumental in nearly every phase of the Citicorp engineering design.

Though each of the architectural concepts that Stubbins created incorporated differing structural constraints, dimensions, and occupancy requirements, Citicorp had made clear, and Stubbins agreed, that certain design criteria must remain constant. First and foremost, the contractual constraints between the bank and St. Peter's Church had to be respected. At least 63 percent of the church's airspace was to remain unencumbered, and there could be no column or other physical connection between the two structures. In accordance with the stated "good neighbor" policy of Citicorp and to take full advantage of available city zoning incentives, a shopping galleria was to be incorporated into each design option. "The provision of small scale shops, boutiques, and restaurants will insure life for the project and is also a corporate statement of responsibility to the public it serves," Stubbins wrote in his proposal.[10]

Each alternative also contemplated the inclusion of a lower-level "sunken" plaza that would link the crosstown IND and the Lexington IRT subway lines, adding a major interchange connection to the city with direct access to the street and the building's elevators.

Finally, the initial plans called for terraced residential apartments at the top of the structure—what Stubbins termed "Sky Terraces"—providing "unparalleled views of the south and north." Stubbins called all of these amenities and requirements the "nucleus" of his proposal.[11]

The bank had given the Medical Chambers cooperative a deadline of October to reach agreement on its critical land component, and thus, finalization of architectural plans was on hold until the acquisition was complete. Though Citicorp still remained noncommittal on the venture despite ongoing meetings with Stubbins and his project architect, Easley Hamner, and vice president of design, Peter Woytuk, behind the scenes, the bank was assembling a team of its own to guide and coordinate the project's development. By the spring of 1972, and in the months to come, Tony Howkins, Robert Dexter, Henry (Hal) DeFord, John S. Reed, and George Herbst joined Henry Muller and William Spencer, all assuming roles in accordance with their expertise. And in May, the bank named George Martin as its initial project manager.

Perhaps sensing the need to provide the bank with further enticement to use the services of an out-of-town office, Stubbins reached out to Richard Roth Sr. of the renowned New York architectural firm Emery Roth & Sons, suggesting a partnership for the architectural phase of the Citicorp project. Family owned and well established in Manhattan, Emery Roth & Sons had become a leader in the design of New York City high-rise office buildings and had often partnered with other firms on projects, including the World Trade Center complex in Lower Manhattan. After several meetings in New York, Richard Roth, perceiving a good fit, agreed to team up with Stubbins as associate architects on the Citicorp Building. In a letter to Roth on June 29, Stubbins wrote, "Yesterday, at a meeting I had with Mr. George Martin, a Vice-President of the Bank, I informed him of this arrangement, and he seem pleased."[12]

Throughout the summer, Stubbins and his associates, Hamner and Woytuk, continued to refine the alternative building proposals and began to acquaint Citicorp's officers with the various personnel required to execute a high-rise building project. From what Stubbins called a "functional organization chart," he introduced prospective engineers and construction managers to his client and spearheaded meetings and interviews to keep the process moving, pending the bank's final decision. By November, Citicorp's officers had interviewed seven construction firms, including Gilbane, Turner, Tishman, Diesel, HRH, Fuller, and Uris, and, more importantly, had finalized acquisition of the city block on which its new building complex would reside.

By the new year, Citicorp was full speed ahead with the project, and Hugh Stubbins, in association with Emery Roth & Sons, had officially—and finally—been named as the project architect. Until then, William LeMessurier had worked exclusively for Stubbins behind the scenes as the engineering consultant, with limited direct contact or discussion with the Citicorp team. As the project shifted into the critical design phase, however, and as technical aspects of the plans began to crystallize, it was clear that engineering services needed to be formally approved by the bank.

Though Stubbins had touted LeMessurier's work and reputation, Citicorp had reserved veto power over the choice of project engineer. Early in 1973, one of the bank's senior officers, Hal DeFord, traveled to Cambridge to meet with and evaluate LeMessurier and his staff. From the start, DeFord made it clear that, despite the engineer's obvious qualifications, the bank would still require his association with another well-known New York structural engineering firm. The Office of James Ruderman, insisted DeFord, had designed dozens of New York office buildings and had often associated with Emery Roth & Sons in the past. The bank would accept LeMessurier's participation in the project but, as a condition to being formally named as its structural engineer, required that he form a joint venture with the Office of James Ruderman.

LeMessurier considered the proposition and agreed to DeFord's terms. "Since LeMessurier Associates already maintained a New York office, sharing the work of jointly producing working drawings was entirely practical," wrote LeMessurier.[13]

"There's an expert on everything in this world," DeFord later said, "and I think we hired them all. We hired consultants to consultants."[14]

In summarizing the central problem faced by engineer and architect in the design of the Citicorp Tower, LeMessurier later recalled, "The building structure from the beginning was a novel and unique solution to the special problem of building a very tall building with the extraordinary constraint of cantilevering out an entire corner over a free-standing church. The unusual structural arrangements were a necessary result of an unprecedented site condition."[15]

During LeMessurier's fateful luncheon with Stubbins at the Acropolis in Cambridge in 1970, he immediately recognized that his "novel and unique" design—four columns positioned midface at each wall of the structure to accommodate the church—though aesthetically innovative, was structurally vulnerable. To counter this, he first contemplated placing a "table" of sorts over the columns and designing a conventionally framed building on top.

"That seemed like the easy way out," he later said wryly. "It was also the most expensive."[16] Then, studying his initial napkin drawing, his mind flashed to the John Hancock Center in Chicago.

LeMessurier's work required technically sound solutions to practical problems. He was professionally curious and an instinctive problem solver. He was also scientifically courageous. As he and Stubbins discussed the peculiar problems posed by St. Peter's Church, he recognized that a novel solution would be required. He recalled that Fazlur Rahman Khan, one of the engineers of the John Hancock Center, had pioneered the concept of "X-bracing" as a way to channel lateral loads on a structure to exterior columns. Khan's groundbreaking design would influence LeMessurier throughout his career, and during his initial meetings with Stubbins, it would inspire a solution to the structural dilemma posed by the Citicorp Building support column placement.

With Khan's work fresh in his mind, LeMessurier began devising a rudimentary series of structurally independent V-shaped "chevrons," each eight stories in height and intersected by mast columns vertically extending the full length of each side of the proposed tower to the base. Through this distinctive use of "super modules," he insisted that wind and gravity loads could be gathered in increments, "like a mother hen and her chickens," and guided down via diagonal compression through the base truss and distributed to the main columns below. Simply put, whatever had been lost by the center placement of the base columns would be gained by this system of bracing support. "This is a very efficient cantilever," he later said. "It is a mathematical ideal."[17]

As Citicorp continued to evaluate the ever-evolving architectural drawings provided by Stubbins and Emery Roth, LeMessurier and James Ruderman began formulating a series of preliminary structural engineering designs for the complex. "The offices of the Joint Venture partners communicated regularly, and computations for the project were made by both parties," wrote LeMessurier.[18]

Throughout the design process, LeMessurier and his lead engineer, Robert McNamara, conducted weekly meetings in New

York with Citicorp's DeFord and Dexter. As the process continued beyond the initial design phase, McNamara assumed most of the firm's tasks, working with Murray Shapiro and Leo Plofker, who had become principals of the Office of James Ruderman after Ruderman's death in 1966.

LeMessurier's innovative chevron bracing system, endorsed by Shapiro and Plofker, was not only structurally efficient but also the most financially feasible choice. Stubbins had provided the bank with a series of sketches illustrating various framing possibilities but made his preference clear. "The chevron bracing is the most economical—and, I think, the best," he wrote to John Reed.[19]

Walter Wriston remained uncertain about the bracing design. Though clearly the least expensive alternative, he worried about its feasibility for the Citicorp Tower. To allay his concerns, Stubbins traveled with the chairman to various other buildings across the country that had employed comparable techniques so he could view, firsthand, how it looked and performed inside and out. The design, Wriston would finally concede, was the only solution to a unique engineering predicament.

By early summer, the parties began to gradually home in on one final architectural scheme, and some of the early proposals were abandoned. A detached core, a structure located outside of the main tower that houses its elevators, staircases, and mechanical systems, one of the highlights of the earlier plans, was replaced by a space- and cost-saving central core. Stubbins and LeMessurier reasoned that the structural symmetry promoted by the central core increased building strength while still meeting tenants' needs. Additionally, this design permitted a novel system of double-deck elevators, further reducing core-to-floor space ratios and enhancing energy efficiency.[20] Simple practicality, however, governed the ultimate decision. "A central core seemed more popular in the New York rental market," wrote Stubbins.[21]

The crown of the building created a fundamental point of contention. Throughout the project, Stubbins had strongly advocated for a distinctive, slanted building top. Various early configurations included a unique multipeaked, sawtooth or scissor-like pinnacle, which soon evolved into a single forty-five-degree slant to accommodate terraced residential apartments—"A bit of Florida beach hotel in the Manhattan sky," according to one observer.[22]

When zoning complications ultimately quashed the apartment proposal, Walter Wriston sensed an opportunity to save money—an estimated $750,000—by abandoning the "Number 1" silhouette in favor of a traditional flat-top configuration. Concerned that the bank's heavy research and development expenses associated with new and experimental automatic teller machines would be an ongoing drag on the company's balance sheet, Wriston and his team approached Stubbins with the downgrade proposal. The architect bristled at the idea. "I held on to the shape because I wanted to capture something of the spirit of the magnificent old skyscrapers—and to get away from the usual boring kind of crate you see nowadays, everywhere," he later said.[23] After all that had been done to ensure the bank's signature identity "by a building of total distinction," snapped Stubbins, he now thought that Wriston was settling. "The decision to build the project with a 'flat top' can be interpreted that the bank has done enough," he wrote to Hal DeFord. "Our conviction is that the uniqueness of the project is in essence denied by building the expedient, or the conventional—a 'flat top.'" He argued that the costs associated with the slant were completely outweighed by the appearance of mediocrity and that since other buildings in the city, such as Empire State, Chrysler, and RCA, were taller, Citicorp must compensate through distinction. "The crown," he argued, "is the element that is lacking."[24] After vigorous debate on the issue, Wriston relented and reportedly told his team, "This guy is the architect, and he thinks [the slanted roof is] the greatest thing. . . . So let's do it."[25] The design would later be christened "Hugh's Masterpiece"—and mockingly assailed as "Walter's Whistle."[26]

Later, as the bank dealt with the potentially disastrous problem of ice and snow sliding from the inclined roof onto the city streets below, one Citicorp official said acerbically, "Why do we always have to be on the cutting edge of everything?"[27]

▲

On July 24, 1973, the bank publicly revealed its chosen final design and plans for construction. The following day, the *New York Times* announced banally, "Plan for Skyscraper on Lexington Ave. Detailed by Citibank."[28] The news barely registered outside of Manhattan. The nation's attention was focused elsewhere.

In dramatic public testimony given earlier that month to the Senate Select Committee on Presidential Campaign Activities investigating the Watergate break-in, Alexander Butterfield, a deputy assistant to President Richard Nixon, exposed the president's use of a clandestine White House taping system, and on July 23, the committee voted without objection to compel release of the tapes. Asserting executive privilege and separation of powers, Nixon rejected the subpoena and barred the committee from accessing the recordings, an act that would lead inexorably to the president's infamous resignation from office.

Despite the dramatic import of national events, final planning and development of the Citicorp Center moved forward undeterred. At 914 feet in height and containing over 1.1 million square feet of usable office space, the Citicorp Tower, once constructed, would become the seventh largest building in New York and the largest bank building in the world.[29]

Housing the financial institution was, however, only one aspect of a dynamic and all-encompassing project. St. Peter's Church, through its stalwart pastor, Ralph Peterson, had separately contracted with Stubbins and LeMessurier for the design of its new church building. Eighty-five feet in height and nestled beneath the northwestern corner of the skyscraper as an independent entity, the church, clothed in Caledonia granite, was to be floated on a bed of neoprene rubber to muzzle the rattle and roar of the nearby Lexington Avenue subway line. A narrow skylight bisecting the full length of the church was included in the plan to provide a prismatic feel to the interior. The majestic centerpiece of the church, the Louise Nevelson Chapel, named after its celebrated designer, was positioned by Stubbins one floor below street level, with large angled windows on Lexington Avenue allowing a view of the sanctuary below. The final church structure was compared to "hands held up in prayer."[30] Stubbins's goal was to integrate the human element into the overall Citicorp Building complex—what he called "a dramatic but friendly confrontation."[31]

Stubbins contemplated a two-level sunken plaza adjacent to the ground floor of the church and encompassing the full base of the complex, which was enabled by the center-faced placement of the tower's nine-story columns. The resulting open space beneath LeMessurier's "stilts" provided an opportunity not seen since Rockefeller Center in the 1930s to use and enhance areas that had been traditionally abandoned or only marginally or superficially developed in the past. Accessed by a wide set of diagonal stairs, the main level of this nine-thousand-square-foot, open-air concourse was intended to provide entry to the subway, the church, the Citicorp Tower lobby, and an atrium. The area was to be decoratively landscaped with trees and outdoor seating and tenanted by various retail spaces and food kiosks. A large fountain with a water sculpture adjoining the stairs was incorporated into the plan to aesthetically conceal the noise and congestion of the city streets above.[32]

"We designed the plaza to bring people out of the subway and into an open space under the building," said Stubbins. "People are then directed up into the light and air through a very broad flight of steps. It's a very enticing idea."[33] "We anticipate that this will be a lively area oriented to the people it serves," he said.[34]

Stubbins and his client further envisaged a low-rise, multilevel shopping galleria with exterior walls sheathed in glass adjacent to the sunken plaza and wedged beneath the Citicorp Tower on the opposite side of St. Peter's Church. "The Market," as it was later called, would be open seven days and feature a variety of upscale restaurants and shops with an international flair. The intent, according to one commentator, was a retail center "functionally distinct from the office tower and other elements of the complex."[35] The interior focus of the galleria was a towering atrium allowing dramatic views of the mammoth rear leg of the skyscraper and revealing what LeMessurier would later describe as "a great deal of scary excitement" at the base of the Citicorp Center.[36]

Throughout the planning phase of the project, countless negotiations and conferences were held between the developers and New York authorities and community associations. The meetings encompassed virtually every topic, from the location of the buildings to the dimensions of the plaza and the type and quantity of commercial and retail activity that would be permitted—and each meeting was conducted with the approval and assistance of the Mayor's Office of Midtown Planning and Development.

◢

Created in 1969 and staffed by a task force of twelve architects from mayor, John Lindsay's Urban Design Group, the Office of Midtown Planning and Development worked with local developers to generate creative ideas in the public interest that would, by 1975, be codified into the New York Zoning Ordinance. Under the measures advocated by the office and the directorship of the architect Jaquelin Robertson, the change in planning philosophy in New York eventually went, according to the noted art and architecture critic Wolf Von Eckardt, "from playing God to seeking God in details."[37]

The design and engineering realities presented by the Citicorp Center allowed for meaningful compliance with the spirit of the task force's work. "It was clear to the Office . . . that at last a project was going ahead that represented their best ideas about what would be good for New York," wrote one journalist.[38] In a prime example of a mutually successful public-private partnership, Easley Hamner, the political mover of Stubbins's architectural firm, worked hand in hand with the planning office staff member Lauren Otis to facilitate all land-use protocols for the project and to coordinate permitting with other public agencies. It was, according to Von Eckardt, "the first public space in New York City to be designed under the Urban Design Group's guidelines."[39]

The bank's voluntary compliance with city planning objectives was, however, according to the *New York Times*, more a function of "enlightened self-interest" than altruistic public service.[40] In return for the incorporation of the so-called privately owned public spaces, or POPS, Citicorp was entitled to build more office space than the zoning law would otherwise permit. This "incentive zoning" allowed Stubbins to design a main office tower with rentable floor area bonuses and a greater vertical height—all to the ultimate profit and value of the bank.

As the regulatory aspects of the project were negotiated, Citicorp finalized its construction team. Despite Stubbins's recommendation of Uris Buildings Corporation as general contractor, the bank chose HRH Construction Corporation, which Stubbins worried did not have sufficient experience with large office buildings. The company, established in 1888, had, however, successfully completed many New York landmark buildings, such as the Beresford and San Remo Apartments on Central Park West and the Whitney Museum of American Art on Madison Avenue. Arthur Nusbaum, HRH's project manager, would notably later fill the same key role on several Trump metro-area projects, including Trump Tower in Midtown Manhattan.[41]

◢

LeMessurier was very proud of the "scary excitement" that he had engineered into the Citicorp Tower. "It's a designer's delight," said Thomas Connolly, project manager for Bethlehem Steel, which was awarded the tower's steel fabrication contract.[42] The Pulitzer Prize–winning journalist Joe Morgenstern later wrote, "The building simply levitates in a graceful way above the church and the plaza."[43] But not everyone appreciated the concept. Even the construction manager had his doubts. "No speculative builder would ever put a building on stilts," Nusbaum told a *New York Times* reporter.[44]

To LeMessurier, however, the building was "fundamentally simple to understand."[45] The so-called trussed tube that housed his diagonal bracing system wrapped around each side of the tower in a series of six structurally independent, eight-story tiers. The system was designed to incrementally transfer the wind and gravity stress loads through five-foot-wide vertical mast columns running down the center of each side of the building to a two-story truss at the base, which in turn distributed the load to the four main columns below. These "super columns," each 24 feet square and 127 feet in height, were bored through the earth to the building foundations, which reached bedrock at an approximate depth of 50 feet.[46] The design, he concluded, was both "economical" and "elegant."[47]

LeMessurier had initially wanted the steel chevron bracing system to be exposed and visible, similar to the John Hancock Center in Chicago, which inspired his design. "I'm very vain," he later said. "I would have liked my stuff to be expressed on the outside of the building."[48] Stubbins, however, vetoed the idea in favor of keeping the building "simple in its skin."[49] There was already enough drama happening at the top and bottom of the structure, he reasoned, and to expose the structural support system would be jolting to the public eye and interruptive of the clean look he was seeking.[50] "He would clearly prefer to entertain us than to frighten us," wrote the architecture critic Paul Goldberger in the *New York Times*.[51]

Indeed, Stubbins chose to cloak the exterior of the building (and The Market) with a skin of natural-colored, refulgent aluminum alternating with bands of reflective glass. This anodized "satiny silver-colored façade" distinguished the center on the city skyline and, according to Stubbins, made "a powerful thrust into the era of space-age architecture."[52] "It glows rather than glares," wrote Goldberger.[53]

The exterior of the building was designed to be not only visually striking but also energy efficient in a time of soaring heating and fuel costs. The windows chosen for the building were insulated, double glazed, and reflective, and the aluminum spandrels were backed by twice the typical thickness of thermo-fiber insulation to reduce energy consumption.[54]

In the midst of the Arab oil embargoes of the early 1970s, Citicorp sought to incorporate other energy efficiencies into its new building. As plans for the design of the structure were revealed, the bank announced, "We intend to put technology to work to reduce our energy requirements."[55] Stubbins later wrote of the subject,

> Baffled low-brightness light fixtures reduce the wattage needed to provide acceptable light levels, which are varied according to intensity of use. Zone control of the tower permits air conditioning to be supplied to specific floors while others are shut off when not needed. The system, which makes maximum use of outside air in good weather conditions, is equipped with special filters to clean recycled air and reduce fresh air intake when conditioning is needed. A tunnel connection with the bank's existing offices at 399 Park Avenue will permit operation of both buildings from a single power plant during offpeak times. The building's management system will be based on computer operation that will automatically match the flow of energy to actual demand. Energy can be stored in the building to reduce reliance on purchased energy in peak periods.[56]

At the early stage of construction, Robert Bell, the director of research and development for Consolidated Edison—and president of St. Peter's Church—approached Stubbins and Joseph Loring, Citicorp's mechanical and electrical contractor, with the innovative and much-touted idea of incorporating solar energy panels into the controversial slanted crown of the building. Studies conducted by MIT and sponsored by the National Science Foundation revealed technological feasibility of the proposal, but when the projected cost savings proved inadequate, the concept was tabled. "The idea . . . and the theory . . . are not enough," wrote Stubbins. "Technology, industry, economics, and psychological acceptance have to be tuned to a fine balance before a concept becomes a solution."[57] Joseph Loring nonetheless proclaimed the Citicorp Center as "state of the art in energy conservation for a high-rise office building with multiple tenant occupancy."[58]

◢

As Stubbins and the Citicorp officers explored energy conservation and other cost-saving channels, LeMessurier and his team began a technical stress test of his innovative structural design. Despite being one of the tallest buildings globally, the Citicorp Tower stood out for its exceptional structural efficiency, resulting in a remarkably low density. At twenty-two pounds per square foot, the unique steel frame of the tower weighed significantly less than those of other comparable buildings but, according to LeMessurier, made the overall structure "more dynamically excitable" than its heavier peers.[59] The strength of the wind-resisting system and the general stability of the tower thus became the engineer's primary concern. "The essential problem," he confirmed, "is to control and limit movement in high winds."[60]

Though rarely affecting structural integrity, wind-related disturbances in high-rise buildings were, at the time of Citicorp's design, a source of human discomfort and economic loss. Liberty Plaza in Lower Manhattan, for example, suffered elevator malfunctions during heavy winds and a "treacherous and humanly unmanageable wind tunnel" in the years following its construction.[61] In Cleveland, violent downdrafts from the thirty-nine-story Erieview Plaza blew passing pedestrians to the pavement, while excessive building sway cracked partitions and nauseated upper-floor occupants of New York's Gulf and Western Building. And the Sears Tower in Chicago and John Hancock Tower in Boston suffered falling glass window panes during strong winds, causing danger to pedestrians, opening delays, and legal turmoil.[62]

The concern continues to this day. Driven by a paucity of buildable land and the willingness of wealthy buyers to pay exorbitant prices for dramatic metropolitan views, many modern skyscrapers, most notably those surrounding New York's Central Park, have become ever narrower and ever taller. In 2015, the *New York Times* wrote, "On a typically breezy day, a tower 1,000 feet tall might move a couple of inches, according to Rowan Williams Davies and Irwin, consulting engineers. About once a year, a 50-mile-per-hour wind comes up, moving a tower of this size about half a foot. On a rare day, say once every 50 years, 100-mile-per-hour winds might move the tower as much as two feet."[63]

To counter the very real problem of excessive building movement and vibration in the ultra-light-weight Citicorp Tower, LeMessurier calculated to his chagrin that he would need to add an additional twenty-eight hundred tons of sheer concrete or steel to enhance the building's stiffness to a degree necessary to ensure human comfort.[64] Recognizing the financial impracticability (and probable ineffectiveness) of such a measure, he began to consider alternative solutions. He sought out the help of the Princeton University professor Robert Scanlan, who had worked with large shock-absorbing devises or "dampers" to limit bridge movement and with whom he had briefly conferenced on a prior project. Considered a leading expert in the field of vibration control in tall structures, Scanlan suggested to LeMessurier the possibility of incorporating a similar damping device into the wind-stabilization design of the Citicorp Building to counter excessive movement.

Damping technology had been used since 1909 for power-line cables, machinery, ships, bridges, and combustion engines but was relatively untested at the time as applied to building sway and oscillation. LeMessurier knew of no such device in a US skyscraper—and Scanlan knew little about its application to buildings. LeMessurier was essentially on his own. His first step was to obtain reliable predictions of wind forces at the Citicorp Center and to determine exactly how the building would perform under those forces with and without the damping system.

To obtain that data, LeMessurier turned to the Boundary Layer Wind Tunnel Laboratory at the University of Western Ontario in Canada. The celebrated founder and director of the center, Alan G. Davenport, had conducted groundbreaking research into real-world atmospheric disturbances in the Earth's so-called boundary layer (between the surface and three thousand feet) and tested those conditions in a six-by-eight-by-one-hundred-foot-long wind tunnel against static and aeroelastic models of specific buildings.[65] The simulation would be dependably used by Davenport throughout his career to evaluate wind dynamics on numerous buildings, such as New York's Twin Towers, the Sears Tower in Chicago, and the CN Tower in Toronto, and would ultimately earn him the moniker "Wind Wizard."[66] To LeMessurier, the process was simply "the most sophisticated wind tunnel test in the world."[67]

After extensive testing, Davenport's simulator established that movement at the upper floors of the Citicorp Building during periods of predicted high wind forces would probably exceed acceptable comfort levels for occupants. He concluded that a damping system would indeed reduce the dynamic response of the building to these conditions. After consultation with his engineering partners, the architects, and the bank's team, LeMessurier decided to incorporate the device into his structural design.

The novel application of a so-called tuned mass damper (TMD) to the Citicorp Tower required complete top-to-bottom formulation and design. LeMessurier's principal engineer on the project, Robert McNamara, took a leading role in the development of the device because he was fully acquainted with the most recent innovations in structural wind resistance. For the more technical aspects and actual fabrication of the system, LeMessurier turned to David Wormley of MIT's Department of Mechanical Engineering and Niels Peterson of MTS Systems Corp. in Minneapolis, a leading manufacturer of high-precision seismic simulators, electrohydraulic testing systems, and shock-testing machinery for military equipment. The final result, said LeMessurier, was "one of the most ingenious devices" he had ever seen.[68]

The central component of the TMD was a twenty-nine-foot-square, eight-foot-thick, four-hundred-ton yellow block of concrete set on a steel plate near the top of the building.[69] Citicorp's first building manager would affectionately refer to it as "that great block of cheese."[70] Based on the principle of inertia, the device, when activated through a series of motion transducers, elevates the concrete block on a thin bed of oil. The free-floating block then shifts in counterbalance to the sway of the building.[71] The movement of the damper can be synchronized or "tuned" to the natural oscillations of the building to enhance the system's response to vibration. "With the movement of the mass thus precisely out of phase with the movement of the tower," wrote one scholar, "the mass will therefore dampen and reduce the motion of the structure."[72]

Stubbins fully agreed with LeMessurier's inclusion of the TMD and touted it to Citicorp as "valid from both engineering and construction viewpoints."[73] But the bank officers remained wary about the possible public-relations impact of such a device. No one wanted to highlight the building's propensity for movement.

After a meeting in New York with LeMessurier and Davenport, John Reed, an MIT-trained executive at Citicorp and the bank's point man for decisions about the TMD, ultimately approved of its use in the tower.[74] Ever cost-conscious and always engaged in design details, the bank's management team nonetheless insisted that each individual component of the device be independently demonstrated in the field prior to installation in the building. "This is *not* a research experiment," quipped one vice president.[75]

While Davenport's Boundary Layer Wind Tunnel Laboratory data enabled predictions of wind motion and force, LeMessurier had repeatedly emphasized to Citicorp that the TMD was designed for comfort of occupants and not for the actual strength and resilience of the building itself. His intricate chevron bracing system, he insisted, was sufficient alone to ensure the structural integrity of the tower under high wind stresses. And the minimum standard for design of that system was governed in all respects by the New York City Building Code.

In the summer of 1973, LeMessurier's Madison Avenue satellite office, staffed by his Cambridge associate Robert McNamara and project engineer Joel Weinstein, began development of the engineering calculations and drawings for the Citicorp Center in consultation with the Office of James Ruderman.[76] The venture worked extremely well, remembered McNamara. Murray Shapiro, the managing partner of Ruderman, though adept at the business side of engineering, was a bit "old school," but the younger McNamara brought a more modern flair to the venture. His innovative use of an IBM 1130 computing system to solve motion equations for the TMD exemplified this "meshing of the old with the new."[77] Otherwise, all load calculations, wind analysis, overturning and resisting moments, calculations for member forces and frame element sizes, as well as connection and lateral bracing strength, for the tower were performed by hand by Joel Weinstein and, to a lesser extent, Murray Shapiro and Leo Plofker.

With the structural integrity of the building fully designed and engineered and with building-code compliance certified by the joint venture, construction of the project was finally set to begin. The old St. Peter's Church and the various other edifices on the block had been adroitly demolished, and on April 1, 1974, beneath a bed of ominous clouds that shrouded Manhattan, Chairman Walter Wriston manned the controls of a bulldozer and ceremoniously broke ground on the Citicorp Center.

Civil engineers retained by HRH Construction and approved by the bank had performed land surveys and geological studies and began excavation of the site, drilling to bedrock in preparation for the reinforced concrete deep foundations—the base of the tower's four columns. Bethlehem Steel had begun fabricating the structural members of the building to the specifications delivered by the engineers, and within a year of groundbreaking, actual erection of the tower began.

In the coming months, with construction of the skyscraper well under way and its distinctive diagonally braced skeleton revealed against the New York City skyline, a reporter for the trade magazine *Architectural Record* wrote, "Even an untutored sidewalk superintendent examining Citicorp Center's unsheathed steel frame perceives that this is something new and different, something exceptional in the way of skyscraper structure. To the structural engineers, the structure represents a clean design 'so simple it can be analyzed by hand,' and the frame, however curious its initial appearance, does possess a straightforward and efficient elegance."[78]

3

"A SKYSCRAPER FOR THE PEOPLE"

With New York nearing bankruptcy and the Arab oil embargoes weighing heavily on the national economy, Citicorp was the only significant office construction project in the city during 1974 and '75. Unemployment among ironworkers in the New York metropolitan area had risen to 20 percent, and plumbers, carpenters, sheet-metal workers, and masons fared even worse. In all, about 40 to 50 percent of the city's one hundred thousand construction tradespeople were out of work.[1] The Citicorp project provided a welcome respite in the economic malaise. Though HRH Construction virtually had its pick of the best subcontractors and workers from each required trade and at rock-bottom prices, the project ultimately employed about 3,000 construction workers, around 550 of whom were on-site at any given time.[2]

On October 6, 1976, a clear and seasonably warm day in Manhattan, the final steel girder of the Citicorp Tower was lowered into place and draped with a US flag and a Citicorp flag beneath it. Walter Wriston told a gathering of three hundred bankers, construction workers, and clergy on Lexington Avenue that the professionals who constructed the tower "are the best in the business," and because of their efforts, "Citicorp promises to become that rare modern commodity: a new building which not only opens for business on time, but which enhances the neighborhood in which it stands." "This new building," he continued, "is a very tangible expression of our faith in New York and New Yorkers."[3]

Not everyone, however, applauded the event or the soaring rhetoric that marked it. Later in the day, at the festive topping-out party on the fifty-first floor of the tower, where refreshments and music enlivened bare lightbulbs and unfinished office space, the construction worker Ronnie Adams commented, only half jokingly, "I wish it would go another 100 floors. I'd be employed another three years."[4]

Several weeks later, at the cornerstone ceremonies for St. Peter's Church, Wriston again spoke poignantly: "This city of ours has suffered of late more than its share of troubles. But they won't last forever. And I have a feeling that when future New Yorkers look back and ask themselves, when did the bad news end in this part of Manhattan and the good news start, it will not be too much of an exaggeration to answer once again . . . that it was at the door to St. Peter's."[5]

And on October 12, 1977, at the dedication of the Citicorp Center, with Governor Hugh Carey and Mayor Abraham Beame in attendance, Hugh Stubbins called his creation "a skyscraper for the people."[6] At a choreographed moment in the ceremony, the fountain in the plaza was activated for the first time, but instead of a calming and artistic cascade of water, the fountain spewed mounds of suds that rolled down the stairs and into the plaza and subway below. A worker had mistakenly dropped several boxes of detergent into the water system. "Absolute chaos," LeMessurier recalled. "It was funny."[7]

Nearly every step of the Citicorp construction process was scrutinized by the press. Even before groundbreaking, the Pulitzer Prize–winning architecture critic and native New Yorker Ada Louise Huxtable, though approving of the center's proposed ground-level amenities, panned the building itself. "As skyscraper design," she wrote, "the Citicorp tower could be called, quite simply, awful. It reaches weakly after image in its ill-advised, sliced-off-and-angled top, arrived at for obvious identity and public relations purposes, in an unlikely combination of disturbing gimmickry and denatured Bauhaus blandness. It has neither romanticism nor structural rationalism but, instead, appears to have been painstakingly invented with a tortured logic through a series of pragmatic and esthetic compromises."[8] "Put this in Cleveland and it would be the eighth wonder of the world," wrote William Reel in the *New York Daily News*. "In New York people hardly look up."[9]

The overall reception of the completed center, however, proved markedly less damning. Robert Mehlman commented in *Industrial Design*, "Besides being esthetically beautiful, Citicorp is probably the most innovative and technologically advanced architectural complex in America."[10] The *New York Times*' Paul Goldberger wrote, "By any standard the architect, Hugh Stubbins & Associates, . . . has created one of New York's significant buildings. . . . Citicorp is a brilliant, dazzling achievement, one that will probably give more pleasure to more New Yorkers than any other high-rise building of the decade."[11]

The praise was not limited to New York writers. In an article titled "Citicorp Center the Toast of New York," the *Boston Globe* architecture writer Robert Campbell lauded the complex as "smasheeroo, no question about it. It's all New York, Broadway-glamorous, showbiz-spectacular. Citicorp reminds you not of the financially crippled and vandalized New York of today but rather of the New York of your childhood. . . . It has that kind of glitter, that indispensable flashiness."[12] The entertainment columnist Jack Egan called Citicorp Center "distinctive and elegant" in the *Washington Post*.[13] And even in London, the building was praised in an article in *The Observer* titled "Transparently Beautiful" as "easily the best New York skyscraper since the great days."[14]

In 1978, the City Club of New York, in recognition of the far-reaching impact of urban design on the metropolitan landscape, granted the Citicorp Center the prestigious Bard Award, which, since 1964, had "traditionally gone to the best, if not the most adventurous, examples of architecture and urban design."[15] In the same year, the American Institute of Steel Construction awarded the skyscraper its Architectural Award of Excellence.[16] And in 1979, the American Institute of Architects recognized the Citicorp Tower with its annual Honor Award, as "a tour-de-force for the skyline as a stylish silhouette; and for the pedestrian a hovering cantilevered hulk."[17]

The Landmarks Preservation Commission of New York acknowledged the Citicorp Center including St. Peter's Church as "a major example of late 20th-century modern architecture" and, because of its "special character and . . . special historical and aesthetic interest and value as part of the development, history, and cultural characteristics of New York City," designated the center as a city landmark in 2016.[18]

Citicorp Center was, perhaps, LeMessurier's crowning achievement. His novel wind-bracing system and innovative application of damping technology earned him international acclaim, and in February 1978, he was inducted into the National Academy of Engineering—one of the highest honors bestowed in the profession.

"You never really know if you're any good or not unless your peers kick you in the ass and tell you so," he later said.[19]

◢

Despite some initial difficulty encountered by Bethlehem Steel in maneuvering into place the base truss sections of the building, some weighing up to fifty-five tons, the Citicorp Tower was completed in a normal fashion, on time, and with little fanfare or further problems.

After LeMessurier and his colleagues completed the structural design of the building, they were essentially uninvolved in the project—with one curious and seemingly innocuous exception. The final bid-set of structural plans drawn and approved by the LeMessurier-Ruderman engineering team called for the connecting joints of all diagonal wind-brace members to be secured with full-penetration welds. "I know that because I saw the drawings finished," confirmed LeMessurier.[20] Seven months after the dedication of the Citicorp Tower, however, while consulting with an architect and a representative of American Bridge Company on another proposed building project in Pittsburgh with a similar wind-bracing system to that used in New York, the men asked LeMessurier how the field connections were made.

"We welded them," he confirmed. Unconvinced that welds were necessary or appropriate for the Pittsburgh project, the American Bridge representative pressed LeMessurier on how the process worked with Citicorp. According to LeMessurier, he telephoned Stanley Goldstein, who, by then, was the managing partner of the firm's New York office, and inquired about the welding process for the tower's chevron brace connections.

"Didn't you know?" responded Goldstein. "Bethlehem Steel . . . offered $250,000 back to the bank if they could redesign a different way of connecting [the braces]."

"We replaced those welds with bolts."[21]

4

DIANE HARTLEY

"From the first he loved Princeton."[1]

Enamored with the institution that accepted him "on trial" and from which he himself would never graduate, F. Scott Fitzgerald, in his first novel, *This Side of Paradise*, mused through the eyes of his restless protagonist of "the Gothic halls and cloisters . . . as they loomed suddenly out of the darkness" and "the silent stretches of green, the quiet halls with an occasional late-burning light . . . and the chastity of the spire."[2] Amory Blaine, wrote the author, fancied the "lazy beauty" of his chosen university, with "its atmosphere of colors and its alluring reputation as the pleasantest country club in America."[3]

Many a Princeton University president would bristle at Fitzgerald's characterization, while many a student confirmed it. Yet amid the undeniable grandeur of its "gothic halls and cloisters," other writers and scholars at Princeton would change the world. In the mid-1970s, however, one Princeton student, a native of the region, was simply trying to find her way through the morass of a grueling course load in a discipline that she feared may not even welcome her.

Born in Schenectady, New York, and raised in Bergen County, New Jersey, Diane Hartley displayed an early fascination with buildings and urban design and an understanding of their scientific relation to mathematics.[4] In 1974, she was admitted to Princeton's School of Architecture and Urban Planning, headed at the time by the renowned architect Dean Robert Geddes.

Surrounded by some of the great architectural minds of the day, Hartley nonetheless began to feel disillusioned and isolated from her chosen field. During the early days of a hands-on design course—a sophomore studio—taught by "the darling of the architecture industry," Michael Graves, Hartley casually observed the other students in the class, all of whom seemed to be thriving, and she suddenly felt out of place—"like a fish out of water," she recalled.[5] Perhaps, she thought, she was not meant to be an architect.

During that period of uncertainty, she enrolled in a survey of engineering course called Structures and the Urban Environment, offered by Professor David Billington. Recognized as the pioneer of the discipline sometimes called "structural art," which promotes artistic expression within the confines of engineering principles, Billington, according to one tribute, "inspired a generation of scholars and redefined the consideration of great engineering works from bridges to buildings."[6] Revered by students and colleagues alike, his classes were among the most popular and highly attended at Princeton. "Billington was magisterial in the lecture hall," said one student, who would later become an MIT professor. "We were all spellbound by the force of his arguments, the breadth of his cultural references, the beauty of his images, and of course, the wit of his jokes. To attend a Billington lecture was to be changed forever."[7] Not only would he inspire Hartley to look at the built environment in a whole new way, but to her, Billington was just a "fabulous guy." She began to think that perhaps she, "a lowly architecture student," as she sardonically referred to herself, could someday become an engineer.[8]

That summer, she worked at Burns and Roe, a consulting engineering firm in New Jersey, where she was surrounded with the language and culture of the profession. Though she handled mostly clerical work for the office, she gained a further appreciation for the field and a basic understanding of its demands. As the position came to an end, her boss, the lead engineer in the company, told her, "Diane, if you haven't ever thought about it, you seem to have a head for engineering—maybe it's something you should look into."[9]

What had seemed like a nebulous and untenable pipe dream soon became a reality. That fall, as she began her junior year at Princeton, she immediately transferred to the School of Civil Engineering and changed her major from architecture to the field that now inspired her professional and academic interests.

The transition was not hailed with welcoming parades. "In the '70s," said Hartley, "'engineering' wasn't part of a guidance counselors' vocabulary for girls."[10] In a male-dominated specialty—and at a university that had only begun enrolling women in 1969—she felt scrutinized and uncertain of her capabilities. She was hesitant to tell people of her major, and she worried that the course load would be overwhelming. At a time when other students were beginning to lighten their academic burdens to pursue independent work or research, Hartley was playing a game of catch-up. She began taking engineering prerequisite courses in her junior year *in addition* to other class requirements—and Princeton did not allow summer school or add-on semesters. The course load had to be completed in four years between September and May.

And, as an engineering student, she was required to write and submit a senior thesis to graduate.

⁂

Efficiency, economy, and elegance: the greatest structural works, reasoned Billington, share these three primary qualities.[11] "In our own age when democratic ideals are continually being challenged by the claims of totalitarian societies," he wrote, "the works of structural art provide evidence that the common life flourishes best when the goals of freedom and discipline are held in balance. The disciplines of structural art are efficiency and economy, and its freedom lies in the potential it offers the individual designer for the expression of a personal style motivated by the conscious aesthetic search for engineering elegance."[12]

Within this context, Billington challenged his students to evaluate significant urban buildings by virtue of their scientific, social, and symbolic contributions to their environment—what he called "The Three Dimensions of Structure." He explained,

> The first, or scientific criterion, essentially comes down to making structures with a minimum of materials and yet with enough resistance to loads and environments so that they will last. This efficiency-endurance analysis is arbitrated by the concern for safety. The second, or social criterion, comprises mainly analyses of costs as compared to the usefulness of the forms by society. Such cost-benefit analyses are set in the context of politics. Finally, the third criterion, the symbolic, consists of studies in appearance, along with a consideration of how elegance can be achieved within the constraints set by the scientific and social criteria.[13]

This three-part paradigm would become the central focus of his scholarship and teachings.

Hartley, a Billington devotee, in the fall semester of her senior year, thought it only natural to ask him to be her senior adviser. Recognizing her commitment to the field, he agreed to it, but time was already running short. Her immediate task was to find an appropriate thesis topic, one both personally achievable and also acceptable to her adviser. She began by researching noteworthy urban buildings and asking friends and fellow students for suggestions. The array of choices seemed staggering, but Billington limited her search to significant structures that enhance rather than detract from, or add nothing to, their surroundings. Eventually an old classmate from the Architecture Department recommended that she take a look at the newly constructed Citicorp Center in Manhattan.

At first glance, she was struck by the elegance of the building, with its novel aluminum sheathing and angled roofline. The use of public-private space at the base of the tower and the application of zoning incentives to increase building height also intrigued her. But it was the inclusion of a TMD, the first of its kind in a

US skyscraper, and, of course, the provocative midface location of the tower's main columns to accommodate St. Peter's Church that captured her scientific mind. Citicorp Center, she decided, was the embodiment of Billington's tripartite paradigm—and the perfect topic for her senior thesis.

In her initial discussions with Billington, Hartley discovered that there had been no real scholarly work on the history of the world's tall skyscrapers. Such a history, Billington argued, would greatly enhance the value of her thesis. Perhaps not fully comprehending the full breadth of such an undertaking, she agreed to include it.

By the early fall, Hartley was immersed in research and still carrying a full course load. Living within the granite bulwark of historic Dod Hall, first occupied by Princetonians in the late 1800s, she could, perhaps, feel the presence of students who had come before her. Had others behind those walls found themselves facing a similar academic plight? Had they succeeded? Would she?

Despite sharing her life at the time with two roommates, a "westie" terrier name Gamine, and a college sweetheart, Hartley remained single-mindedly absorbed in her work. It seemed like the world was passing her by. Indeed Elvis Presley, the "swivel-hipped prophet of rock," had died that summer, and a dreaded serial killer, David Berkowitz, the infamous "Son of Sam," had finally been apprehended after a yearlong killing spree that left six people dead and seven wounded on the streets of New York City.[14] She hardly noticed. Her focus was entirely on her thesis.

Through all hours of the night, Hartley sat awake on her narrow dormitory bed awash in three-by-five index cards, charting, researching, and organizing. Sometimes she sprawled her work across a sofa or desk in the study room of Campus Club, an informal, uniquely Princeton dining and social facility, where, on radios down the hall, Lindsey Buckingham and Christine McVie could often be heard crooning, "Don't stop thinking about tomorrow."

From countless books, articles, studies, and drawings, she painstakingly transcribed volumes of information to the index cards and later collated them into systematic outlines grouped by topic.

At first, she converted these into handwritten section drafts to be typed later, but as time became scarcer, she worked directly from the research cards and simply composed the final version on her Olivetti manual typewriter as she went along, concealing the inevitable errors with correction paper.

The thesis, ultimately titled "Implications of a Major Urban Office Complex: The Scientific, Social, and Symbolic Meanings of Citicorp Center, New York City," would contain four distinct sections, one dedicated to the history of high-rise buildings and the remaining three to each pillar of Billington's paradigm. "Since the development of the skyscraper in the late nineteenth century," she began, "this structural form has embodied the essence of urban modernity."[15] The last four words of the sentence struck her as pompously hyperbolic, and with her best friend looking over her shoulder, they laughed loudly as she typed them.

Hartley quickly realized that research books and scientific journals would take her only so far, and so she began writing letters to key members of the Citicorp project team requesting information. In a day when building safety did not emphasize prevention of espionage or terrorist attack, the response was enthusiastic. The bank's officers invited her to their Park Avenue headquarters and cheerfully shared their resources about the construction process.[16] Later, she visited the Citicorp Center on her own, marveling at LeMessurier's "scary excitement," and attended a service at St. Peter's Church, taking in the full flavor of the complex's cultural import. Through an engineering class at Princeton, she attended a tour and demonstration of the TMD guided by Princeton professor Robert Scanlan, one of LeMessurier's original consultants on damping devises.

By the new year, she had also reached out to LeMessurier's headquarters in Cambridge, informing the office of her project and requesting guidance. Her inquiry was quickly passed on to Stanley Goldstein, then the firm's managing partner in New York, and Joel Weinstein, one of the Citicorp project engineers who had performed many of the engineering calculations for the building.

Goldstein struck the undergraduate as a charismatic, larger-than-life figure, and he immediately extended his invitation to the company's Madison Avenue office. Braving icy city winds, she traveled to New York in early February and met with the two engineers who graciously unlocked their vault of information about the Citicorp Tower. "I left their offices with copies of calculations, full-size set of structural blueprints, and went back to Princeton to continue my work," she recalled.[17]

In the ensuing weeks and months, she worked ceaselessly on her project. Since she still needed to reserve time for copying and binding, she had decided to place much of the scientific and technical analysis of the building in a series of appendixes that could be later added to the end of the thesis. Though she completed modeling of the building's dynamic response to wind loads with and without the TMD—what she viewed as the "sexy" part of the project—final exams were fast approaching, and she decided to delay consideration of static winds and dead loads, thinking that would be a less complicated task.

By mid-March, with the thesis due date already extended, Hartley began the remaining technical analysis of the structure. Her method of examining the wind-resistance capabilities of the Citicorp Tower was to break down and replicate the various design-to stresses for the building and to compare that analysis to the actual design calculations provided by LeMessurier's office and noted on the actual structural drawings. She began by assessing the gravity loads of the building—its sheer weight and density—as provided by the engineers, and examining resistance to static wind pressures on the basis of accepted predictions at different heights of the building.[18] In comparing her own calculations for horizontal shear, deflection at the tower tip, and overturning moment, after factoring in the gravity resistance loads, her figures matched very closely with those provided by Joel Weinstein—assuming winds striking only a single face of the tower at a perpendicular angle.

Perhaps because lighter-weight urban buildings with complex structural designs were a fairly new phenomenon in the mid-1970s, the New York City Building Code did not expressly require consideration of nonperpendicular winds in high-rise construction projects. "Consideration of wind from non-perpendicular directions on ordinary rectangular buildings is generally not discussed in the literature or in the classroom," wrote LeMessurier in 1978.[19] Even as an undergraduate engineering student, however, Hartley had been taught and intuitively understood that structures must be designed to withstand not just best-case stress predictions but also more extreme pressures and on their most vulnerable points, especially in view of the new, lighter materials and design and construction techniques that enabled modern architectural innovation.

In Billington's engineering classes, she had studied the Tacoma Narrows Bridge collapse of 1940, which had demonstrated in tragic fashion the need for wind and aerodynamic analysis in bridge design. She assumed a similar investigation was required in an examination of the Citicorp Tower. "The situation of a quartering wind must be examined in addition to that of wind striking full-face," she wrote in her thesis.[20] Given the unique structural design employed by the building's engineers, she concluded, "In a building such as the Citicorp tower, it is crucial to study the effects of a cornering wind, wherein a greater contributary [sic] area of the building face is affected."[21]

Originally observed by mariners on the open sea, so-called quartering, cornering, or diagonal winds in an urban context are those that strike the corners of an edifice, thereby impacting multiple faces simultaneously rather than only one. While perpendicular winds typically impact one face of a structure straight on, quartering winds, by definition, strike at a forty-five-degree angle to the two affected surfaces.

On the basis of the square design plan of the Citicorp Tower, Hartley calculated through "simple geometric relationships" that the net wind load on any given face of a structure from a quartering wind would be 0.71 of the expected force of an identical

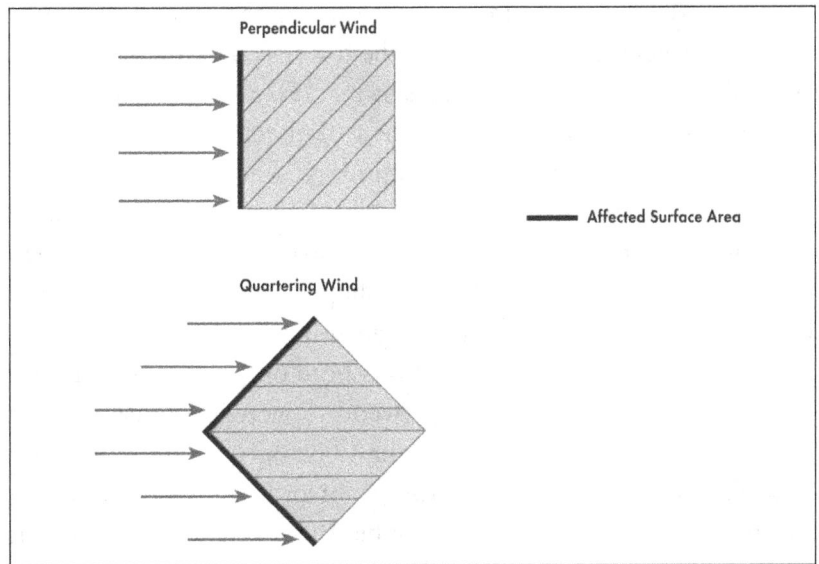

FIGURE 4.1. Diagram illustrating the surface areas of a structure affected by perpendicular and quartering winds.

perpendicular wind (since, of course, the stresses are not applied directly but rather at an angle). However, because quartering winds are simultaneously directed against *two* sides of a building, the total wind load is necessarily doubled. "Therefore," wrote Hartley, "a multiplication factor of $(.71)(2) = 1.42$ is in effect in a cornering wind condition as compared to a perpendicular-face wind condition."[22] Simply put, a conventional building, when subjected to quartering winds, incurs stresses 42 percent higher than in perpendicular winds.

Hartley scoured the construction drawings and load calculations provided by the engineers for confirmation. Surely, she thought, given the novel design of the Citicorp Tower, they would have incorporated an analysis of quartering winds and their effect on the columns, but she could find no evidence that they had. She was convinced that these winds produced the highest stress levels, and yet the official engineering documents seemed to reveal load calculations significantly *lower* than what she thought they should be.

Would LeMessurier's team not have utilized the highest and most conservative design-to stresses for the tower's columns and bracing system? "It just didn't make sense to me," Hartley later said.[23]

She assumed that she had made a mathematical error or had somehow applied an incorrect standard. She worked the numbers over and over, and yet she kept coming up with the same disconcerting result. Though the construction plans had been stamped and approved by Murray Shapiro at the Office of James Ruderman, she noticed that the engineering sheets containing Joel Weinstein's handwritten calculations bore only his initials. Boxes labeled "Checked by" and "Approved by" were left conspicuously blank. Perhaps, she thought, she had been given preliminary figures. She considered bringing the problem to Billington, but she was down to the final two pages of her thesis and was running out of time. She needed to go straight to the source.

◢

Oblivious to wind stresses and engineering calculations, the leadership of Citicorp was focused only on the bold corporate statement that its new signature headquarters projected to New York City and around the world. In keeping with its commitment to remain a vibrant member of the community, the bank hired a talented young woman from Flatbush named Vivian Longo as its building manager, with a staff of over one hundred people. "Aside from being 26 and beautiful," wrote the *New York Daily News*, "Miss Longo owns a master's degree from Fordham, a serious talent for poetry, sculpture, and the piano and the guitar, and a disposition that could bring calm to a rush hour subway car."[24] Through the company's people and appearance, Wriston clearly wanted to promote its exciting, energetic persona, and its officers searched for innovative ways to share that persona.

Citicorp management was very enthusiastic about Hartley's thesis. During her visit to their offices the previous fall, it seemed that the bank's team was as interested in her work as she was in their building. Privately, they viewed her project as an opportunity to further showcase their tour de force on the city skyline. Within days of her visit, the bank made an astonishing proposal. For the publishing rights to her thesis, converted, perhaps, to a coffee-table book or the like, they would pay her a stipend of $10,000, pending review of the final completed document.

She was stunned. Publication was one of the greatest honors a student could receive. And at the time, the starting annual salary of a first-year structural engineer was certainly no greater than the amount she was being offered.

"I was all in," said Hartley.[25]

⸝

"What am I doing wrong?" Hartley asked Weinstein. "Why don't my numbers match?"[26]

She was hesitant to have even made the call. LeMessurier's office was internationally renowned and enormously respected in the field. The Citicorp Tower was observably sound and of worldwide architectural acclaim. She was merely a student with little experience—and even less confidence—in the field. This matter of quartering winds was only a tiny part of her overall thesis, and she had convinced herself that there was definitely a plausible explanation for the discrepancy.

She nervously told Weinstein about her calculations and asked if, perhaps, she had overlooked his analysis of quartering winds. She explained that her figures for perpendicular wind loads accurately replicated with his, but for diagonal winds, the design-to stresses were much higher and, crucially, not reflected in the construction plans, as far as she could tell. Her primary concern was the location of the main columns and their potential vulnerability to quartering winds.

Weinstein patiently listened to Hartley's reasoning and then gently explained that, far from creating a structural hazard in quartering winds, the midface configuration of the primary columns actually enhanced building resiliency rather than detracted from it, as she had feared. With the typical placement of columns at the corners, he insisted, direct wind resistance is provided by only one of the four columns in a quartering wind, while the leg at the opposite corner acts in compression. The remaining two columns serve simply as pivots. With the Citicorp Tower, however, *two* legs are working in tension, and the remaining two are working in compression. In essence, he explained, while quartering winds may indeed generate increased loads, the unique column placement of Citicorp actually produced more efficient resistance to those loads. Because of this analysis, he said, the effect of quartering winds was simply not critical to a practical evaluation of the tower's structural integrity.

Though somewhat perplexed, Hartley was glad for the engineer's explanation. By this point in her senior year, she was utterly exhausted and emotionally drained. She just wanted to complete her thesis and graduate from college. She thus unreservedly accepted his conclusions and incorporated them into her work without protest. "With the Citicorp tower mid-face leg positioning," she wrote, "more effective resistance is provided against a cornering wind."[27]

On Monday, April 24, 1978, Hartley submitted her completed thesis to Billington. In what would ultimately be recognized as a remarkable contribution to the literature of high-rise structures and the study of the Citicorp Tower, the document comprised 460 pages sourced with 440 endnotes, accompanied by ninety-six illustrative plates and twenty-nine mostly hand-drawn technical figures, in two leather-bound volumes. It was, according to Billington, "an outstanding thesis."[28]

The previous Friday, after leaving the thesis at the printer for copying and binding, Hartley returned to her dorm room and fell into a deep and restorative sleep, from which she did not awake until Sunday morning.

For weeks after Hartley's graduation from Princeton, she repeatedly wrote and telephoned Citicorp to inquire about their promise of publication.

She never got a response.

Weinstein was not alone in his sentiments regarding the efficiency of the tower's main columns; the opinion was shared throughout LeMessurier's office. In June 1976, Stanley Goldstein told a writer for *Engineering News-Record*, "With the legs at the midpoint of the faces, in a quartering wind they are all working."[29] And Robert McNamara, lead engineer and, later, principal and vice president of LeMessurier Associates, was quoted as saying, "Bill LeMessurier . . . concluded that the quartering wind did not govern the design and need not be further considered."[30]

Nevertheless, after Hartley had reviewed Billington's comments on her thesis, she would later surmise what she considered to be a disturbing problem with the wind analysis performed by the Citicorp Tower engineers. It is elemental engineering, she would theorize, that the ability of a structure's primary columns to resist a wind force is partly a function of the distance those columns are from the tower's overturning axis, that is, the critical line of inflection around which it would topple in the event of an overwhelming wind or other opposing force. Applying that axiom to the Citicorp Tower, Hartley reasoned that although there may have been two columns in active resistance to quartering winds, those columns, given their midface placement, were located closer to the overturning axis and, therefore, should have been engineered for much greater strength than they actually were.

David Billington put it more succinctly. In his list of comments on Hartley's thesis, regarding her adoption of Weinstein's explanation of quartering wind resistance, he wrote, "Why is it more efficient when column area is doubled?"[31]

Hartley was correct in her assertion that quartering winds presented more variable stress levels on portions of the tower than perpendicular winds did and that the Citicorp engineering team had, in fact, failed to properly analyze their full impact on every aspect of the building. In reality, however, the location of the columns presented no elevated threat to the integrity of the structure from quartering wind loads.

In Stanley Goldstein's statement to *Engineering News-Record*, he pointed out in relation to the center placement of the columns that "the most critical condition is wind hitting a corner, striking two faces at once."[32] If anything, simple perception suggests, ironically, that the columns are less efficient against winds from the *perpendicular* direction since there is only one column in resistance to those winds. Though the midface placement of the Citicorp Tower main columns was innovative in its conception, it had been studied in detail and founded on sound engineering principles to withstand wind loads from all directions. In quartering winds, the tower's columns mathematically afford equal resistance to columns conventionally placed at the corners, and that resistance is more efficiently delivered because all four of the columns are working— two in tension and two in compression. Their proximity to the overturning axis is, therefore, counteracted by increased column mass and engineered strength. Though the designers had indeed failed to adequately analyze quartering winds in every aspect of the tower's wind-bracing design, the bottom columns were nonetheless strong enough to endure them.

Still, unbeknownst to Hartley, the building contained a subtle and overlooked construction feature that resulted in a peculiar and menacing amplification of those wind loads. The Citicorp Tower was indeed in grave danger but for reasons that neither she nor LeMessurier and his team of engineers fully understood at the time.

5

THE MYSTERY STUDENT

By the spring of 1978, Citicorp Center was virtually complete and about one-half occupied. Twenty-six of the tower's fifty-nine stories were earmarked for the bank's growing workforce, while commitments from Ernst & Ernst, Price Waterhouse, IBM, and others were pending for the remaining floors. Pastor Peterson would later report a rise in membership of 18 percent per year after consecration of the newly constructed St. Peter's Church.[1]

Walter Wriston continued to "invest" in the community and at the same time protect his $190 million asset. The "topless neighbors," as the *New York Daily News* called them, such as the Bottomless Pit and Jax Three Ring Circus, housed across the street in several deteriorating tenements that survived construction of the Citicorp Center, irked the chairman, who reportedly told his ministers, "I want those gone."[2] Acting on Wriston's vision, the bank eventually gentrified the area and made a profit in the endeavor.

The office building was, by all accounts, a safe and pleasant workspace. "The performance of the building was excellent," LeMessurier reported. "The tower was perceived by the owner to be unusually rigid in the few high winds experienced, and no plagues of glass breakage occurred."[3]

After completion of the complex, the bank took out a series of full-page advertisements in the New York papers, with a photo looking up at the tower from the massive columns below. "Why is Citibank staying in New York?" read the ad. "We grew up here."

As Wriston's biographer wrote, "While Citicorp Center was not conceived as a symbol of the bank, it quickly became one."[4]

◢

Several weeks after Hartley's last conversation with Joel Weinstein, LeMessurier ironically received a telephone call from a person whom he later described as "an architectural student in New Jersey whose teacher had challenged him to study Citicorp Center."[5] LeMessurier never identified the student by name, nor did he recall the school that the student attended. He did remember, however, that the call came during an informal meeting with his staff—and that the student was male.

At the time, LeMessurier taught a course on structural engineering at the Harvard Graduate School of Design and, ever the educator, was apparently willing to pause his meeting to briefly entertain the student's inquiry. Perhaps LeMessurier took the call thinking that it was Diane Hartley, after he was advised of her earlier communications with the New York office. Whatever the circumstance, years later, an engineer in the firm who was present at the meeting distinctly recalled LeMessurier's secretary opening the door of the conference room and announcing that a "student from New Jersey" was on the phone and that LeMessurier left the room to take the call.

According to LeMessurier, the student said he was writing a paper on the Citicorp Center and that he called to inquire about the tower's novel engineering features and the unusual placement of the four main columns. His professor had told him that the engineers had located the columns in the wrong place, that it would have been safer to position them at the corners like other conventional buildings.

"Listen," snapped LeMessurier, "I want you to tell your teacher that he doesn't know what the hell he's talking about, because he doesn't know the problem that had to be solved."[6] LeMessurier told the student that he had to return to his meeting but vowed to call back to clarify.

About half an hour later, to the student's amazement, the phone rang. As promised, the engineer explained, more patiently, that the midface placement of the columns was necessitated by the unique site challenges presented by St. Peter's Church. He informed the student that the location of the columns and the distinctive geometry of the structure itself created even greater resistance to the punishment of diagonal wind loads—a concern that the student had not even mentioned and was completely unaware of. The columns as designed, LeMessurier assured him, were much more efficient than his teacher suggested. "Now you really have something on your professor, because you can explain all of this to him yourself," he boasted.[7]

Though Diane Hartley never claimed to have spoken directly with LeMessurier and though the student's call was apparently made weeks after she turned in her thesis and moved on with her life, nearly everyone expressing interest in the matter in later years concluded that *she* was the student who made that call. LeMessurier had at various times interchangeably referred to the caller as a "student of engineering"—which Hartley was—rather than an "architectural student," and it was assumed that he was either mistaken as to the gender of the caller (he always referred to the student as "he") or that he confused a call from Weinstein or, perhaps, Goldstein relating Hartley's concerns as coming from the student personally.[8] And either scenario may, in fact, have been true.

In November 2011, however, a New Jersey architect by the name of Lee DeCarolis sent an email to LeMessurier's firm in Cambridge explaining that he had recently learned of a controversy with the Citicorp Tower design and of the mysterious call from a student in New Jersey. "It turns out that I am that student and remember the conversation pretty well," he wrote. "When I read the account it gave me a shudder," he continued.[9]

Later, DeCarolis wrote an article for the Online Ethics Center for Engineering and Science publicly identifying himself as the mystery student. He claimed that he learned in 2011 of some controversy with the Citicorp Building from a book authored by Steve Silverman called *Einstein's Refrigerator and Other Stories from the Flip Side of History* and that he later informed LeMessurier's engineering firm that it was he who made the call to its founder three decades earlier.

DeCarolis wrote that in 1978, during the second semester of his freshman year as an architecture student at the New Jersey Institute of Technology in Newark, New Jersey, he had taken a class on the basics of structural engineering, and the professor, John Zoldos, had asked the students to write a paper about a building with an unusual structural feature. DeCarolis had read about the newly constructed Citicorp Center and, since he lived near the area, decided to make it the subject of his report. Zoldos was familiar with the building and had noted some concerns about its design. "The details are a little hazy as I try to recall," wrote DeCarolis, "but Professor Zoldos expressed some reservations about the building":

> He thought it might have been overly ambitious in its attempt to create a cutting-edge design. In particular, Professor Zoldos mentioned the columns being located not at the corners but in the middle of the sides of the building. I remember him saying that the first job of a structural engineer is to design a safe building. I was surprised by his criticism of this great, new, and much-admired building. I was a very naïve student at the time and thought that any architect that designed such an amazing building was beyond criticism.
>
> I decided to call the architect, Hugh Stubbins. His office suggested that I call the structural engineer instead. So I called William LeMessurier in Cambridge, Massachusetts, and found myself very impressed to be talking with the structural engineer of the CitiCorp Building. We talked about the building and its perilous columns.

LeMessurier gave me lots of information, but after a few minutes, said he had to go to a meeting and would call back. I did not think he would, but about a half hour later, he did. I was stunned.

We talked for another 15 minutes or so, discussing the columns and their location. Looking back, I must have seemed hopelessly ignorant to LeMessurier, telling him that my professor thought the building columns were in the wrong places. I also asked about the tuned mass damper I'd read about. It was very hard for a freshman architecture student to understand how it worked and realize what a technological innovation it was.

That's the full extent of my involvement with the Citicorp Building. I didn't tell LeMessurier there was anything wrong with his building. I knew nothing about quartering wind loads or the steel frame being bolted instead of welded. The Citicorp paper and the grade it received have receded in my memory for the following 33 years.[10]

It is impossible to confirm DeCarolis's claims with certainty. His paper no longer exists, and he has no notes or other evidence of his conversations with LeMessurier. He admittedly did not question the design or stability of the building; he simply conveyed to the engineer what his professor had told him. LeMessurier insisted that he received the call in June, about a month after he learned of the bolted bracing connections. DeCarolis thought he made the call earlier, perhaps in April or May, prior to the end of his semester. Sometime in the spring of 1978, however, LeMessurier *did* receive a call from an architectural or engineering student in New Jersey who questioned the placement of the Citicorp Tower's main columns, per his professor's suggestion. Diane Hartley did not make that call, and notably, no other "student" has come forward claiming to have spoken to LeMessurier or anyone else in his office about the Citicorp Center.

Hartley, on the other hand, had conducted extensive research on the building, and she wrote a senior thesis analyzing its structural features. She was the person who addressed in a rudimentary way a potential vulnerability of the tower to quartering winds, and she had extensive conversations with the firm's engineers about her findings. Though she never spoke directly with LeMessurier, there is every reason to believe that Joel Weinstein may have passed her concerns up the company chain of command. Weinstein later claimed little memory of the content of his discussions with Hartley but confirmed that if he had been confronted with building safety issues or potential engineering miscalculations of the type she raised, he would have brought those up with LeMessurier or other superiors in the office.[11]

Though neither Hartley nor DeCarolis definitively claims to have exclusively influenced LeMessurier's actions, there is little doubt that each, to some degree, profoundly impacted what would happen next.

6

"SOME VERY PECULIAR BEHAVIOR"

Several months after the groundbreaking ceremonies for Citicorp Center, as the Bethlehem Steel design engineers began formulating shop drawings for the steel members and splice connections of the tower's chevron bracing system, they immediately recognized that the requirement of full-penetration welds in the LeMessurier-Ruderman bid design plans was excessively cautious, overly expensive, and needlessly time-consuming. Despite the sagging economy, specialized and proficient welders were not easy to find, and a welding process would delay and unnecessarily prolong construction. Consequently, Bethlehem suggested the more economical and equally safe use of bolted splices in place of welds and offered the bank a financial incentive if allowed to make the change.

On August 1, 1974, after conducting analysis of the proposal and calculations of tensile capacities for the new bolted connections, LeMessurier's New York office, on behalf of the joint venture, approved the substitution and provided the fabricator with the forces necessary for the bolt design. High-capacity bolts manufactured to the engineers' specifications were thus incorporated into the tower's bracing system in place of full-penetration welds.

Whether LeMessurier personally knew about the change is unclear. He would later claim that he was unaware until 1978, when he says it was brought to his attention while meeting an architect and construction managers on another project in Pittsburg. Robert McNamara would insist, however, that LeMessurier knew of the change at the time it was made and approved it.[1]

Regardless, LeMessurier stated many years later what most structural engineers would certainly have concluded about the change: "It was a perfectly reasonable request," he said.[2]

And under ordinary circumstances, he would have been right.

⊿

LeMessurier could not get his conversation with the New Jersey student out of his head. At the time of DeCarolis's call, LeMessurier had been working through the critical wind directions of a triangular building in Boston, most likely the Back Bay Hilton Hotel, and his mind had already been focused on wind stresses and effects. The question of quartering winds and the column placement of the Citicorp Tower was intriguing. He began thinking about the structural engineering class that he taught at Harvard and, later that day, decided to prepare some notes for a fall lecture on the "unusual geometry of Citicorp's column and wind bracing system."[3] He was confident it would make for a fascinating and informative study topic. "I always dig into things like this. . . . I'm a teacher myself," he told one commentator.[4]

With Diane Hartley's broad assertions perhaps vexing LeMessurier, he pondered the effect of quartering winds, not just on the four main base columns but on *every* aspect of the structure. Hartley was not aware of the change from welded connections to bolts, and her conversations with Weinstein, even if conveyed to LeMessurier, had not emphasized the possible nuanced effects of winds on any particular elements of the building beyond the bottom columns.

In the spirit of what LeMessurier called "intellectual play," he began formulating some drawings and hand calculations comparing the impact of perpendicular and diagonal wind forces on the

Citicorp Tower.[5] Though he was able to confirm that the location of the bottom columns presented no structural concerns, he intriguingly uncovered what he later called "some very peculiar behavior" with the tower's chevron wind-bracing system.[6] He determined that in a quartering wind, the stresses in four of the eight bracing members in each module vanished to zero—but they *increased* by 40 percent in the remaining four.

The Citicorp Tower had been designed in such a way that its diagonal braces were preloaded in compression. In other words, these members were squeezed or pressed together due to the gravitational weight of the building. But the braces were also tasked with transmitting wind loads to the columns below. When the wind blew directly against the structure in a perpendicular direction, these diagonal braces remain compressed due to the combination of gravity and wind tension. When the wind blew at an angle, however, certain of those braces partially unloaded from their compressed state in such a way that amplified the tension on them.

LeMessurier thus discovered that the diagonal braces bearing both compression and tension were affected differently by winds depending on their direction. The change in wind from perpendicular to cornering would cause an unloading of the compressive force on these braces in a manner that resulted in an unexpected increase in tension. "I should have known better," he later said.[7] "I could have just thought about it and never would have gone through it in the dumb way that I did."[8]

Now his mind flashed to the bolted brace connections.

If the overall chevron bracing system was indeed reactive to quartering winds, then, he feared, so too must be the splices that held that system together. He knew that had the original specifications not been changed from full penetration welds for those splices, the bracing system would have provided more than ample resistance to winds from any direction, even with the 40 percent increase in tension inherent in some of the members.[9] But with that change, the strength and capacity of the bolts that secured the connections instantly became a matter of unease.

LeMessurier was all but certain that his engineers in New York had not considered the amplified forces from diagonal winds when formulating the load capacities of the bolts for Bethlehem Steel. Indeed, they had adopted what LeMessurier insisted was an industry-wide interpretation of the building code that required structural design for perpendicular winds only. "It would be a miracle if they ever thought . . . about diagonal winds," he later recounted.[10] "I didn't go into a panic over it, but I was haunted by a hunch that it was something I'd better look into."[11]

◢

"'I would like to thank'—here he started breaking down and crying—'the Yankee management, . . . the press, the news media, my coaches, my players, . . . and most all'—he began crying more heavily and could not speak for nearly 10 seconds, then gasped out softly,—'the fans.'" In what would be the first of several dramatic departures for the "tempestuous street fighter," as the *New York Times* called him, Yankee manager Billy Martin announced his resignation on the balcony of an antique shop in Kansas City, where the Yankees had arrived for a two-game series against the Royals. The departure was inevitable. Referring to the team's slugger, Reggie Jackson, and its owner, George Steinbrenner, Martin inanely told a reporter the night before, "The two of them deserve each other. One's a born liar, the other's convicted."[12]

On July 24, 1978, New Yorkers, unaware of quartering wind loads or the structural peculiarities of their skyscrapers, were abuzz only with Martin's tearful good-bye. Like the unspoken trust between old friends, the safety of their buildings was a given.

That morning, LeMessurier flew to the city to meet with his New York engineering team. He was not overly troubled about the strength of the bracing system; it had been designed to withstand stress levels in excess of what he had assumed in his calculations. "Forty per cent sounds like a lot to a layman," LeMessurier said, "but it doesn't sound like imminent collapse. . . . We have factors of safety generally speaking much higher than that."[13] At the same time, he felt that he could not just ignore the issue. "I'd rather be put at ease," he said.[14]

LeMessurier had previously asked Stanley Goldstein to gather all the shop drawings for the Citicorp Tower splice designs, most of which were housed at Ruderman's office, so he could study the connections in detail during his visit. With the plans draped across the conference room table, he asked Joel Weinstein to dig out the calculations used to determine the splicing forces relied on by Bethlehem Steel in fabricating the bolts in the summer of 1974.

As LeMessurier anticipated, the documents affirmed that his engineers, trained and practiced in conventional high-rise structures, had indeed considered only perpendicular winds in the design of the bolts. "This design tradition and mental set operated to foreclose inquiry by any member of the team into the significance of 45° wind forces," insisted LeMessurier. "And, though Stanley Goldstein had analysed the effect of 45° winds on the four principal columns, since the effect of such a wind on the columns was *not* critical, it seems likely that intuition was misled into feeling that *no* element of the building was sensitive to 45° winds."[15] As LeMessurier further examined the figures, however, another "subtle conceptual error" emerged.[16]

Since, as LeMessurier knew, high-rise buildings are often subjected to enormous levels of wind tension, they are designed to resist that tension, to some degree, through offsetting gravity compression. Simply put, the sheer mass and weight of the structure and its contents tends to hold the building together in the face of wind loads. Though factors of safety are required by code to be incorporated into the design, tension caused by wind can often exceed compression by gravity, and thus, in the case of the Citicorp Tower, the splicing bolts needed to be sufficiently engineered to withstand that difference.

To LeMessurier's dismay, however, the calculations performed by his New York engineers revealed that rather than deducting 75 percent of the permanent gravity load in the building to offset wind tension, as required by the American Institute of Steel

Construction Specifications, they deducted the *full* compression, according to LeMessurier.[17] In so doing, they concluded that only "a practical minimum of four bolts per splice" were required, each bolt having a diameter of around an inch and a half.[18] "Now if the effect of 45° wind is considered," LeMessurier advanced, "the gross wind tension will increase by 40%." (This was essentially Diane Hartley's assessment of quartering winds.) In that case, LeMessurier deduced, instead of four bolts, more than ten bolts would have been required per splice. And, had his team properly deducted only 75 percent of the gravity load to offset quartering wind tension rather than 100 percent, fourteen bolts would have been needed. "In a situation where gravity load offsets wind tension there is enormous leverage on the effect of an increased wind force," he added. "If the gross wind force increases by 40% the net wind for bolt design increases by 160%."[19]

LeMessurier was now forced to conclude that all previous margins of error incorporated into the Citicorp Tower design—so-called factors of safety—had now vanished.

"This thing is in real trouble," he began to think.[20]

7

"VERTICAL URBANISM"

"He was a beautiful soul," his mother said.[1]
Though Christian Regenhard could have chosen
any other vocation, he followed in his father's
footsteps and joined the Marine Corp. A graduate of the elite Bronx High School of Science,
Regenhard was exceptionally intelligent and
exquisitely artistic. He enlisted on his nineteenth
birthday, and after five years of distinguished
service that earned him twelve medals and honors for excellence, he left the Corp seeking new
journeys and experiences. He spent some time at
San Francisco State University studying art and
honing his writing skills but much preferred the
adventures of travel, rock climbing, scuba diving,
and, as his mother remembered, breaking hearts
with his charm and uncommon good looks.

By January 2001, Regenhard was back home
in the Bronx and, again following his father's
lead of public service, decided to become a firefighter. He graduated from the New York City
Fire Academy on July 27, 2001, and was assigned to Ladder 131, Engine 279, Red Hook
Firehouse, in Brooklyn as a probationary recruit.
He was twenty-eight years old, good-humored,
and socially gifted. His brothers in service agreed
that they had never met anyone like Christian
Regenhard.

On September 11, 2001, Regenhard's company
responded to the burning Twin Towers of New

York's World Trade Center, and though he had only been on the job for six weeks, he heroically entered the buildings in an effort to rescue trapped civilians. At 10:28 a.m., as Regenhard was assisting a fellow firefighter in the lobby, the North Tower collapsed. He was never seen again.

From Regenhard's bedroom in his parent's Co-op City apartment, Sally Regenhard, Christian's mother, peered out the window to the unfolding horror in Lower Manhattan. With disbelieving eyes, she watched as the towers burned and then disappeared into an ashen plume. She realized that her son was there, though she refused to accept the possibility that he was gone.[2]

In the weeks following the atrocity of 9/11, Sally hired a detective to search for her son in the local hospitals. She and her family constructed missing-person bills and posted them across the city, as thousands of others had desperately done. As the hours turned into days, and still with no word, she told herself that he was, perhaps, still alive beneath the rubble. Surely, she thought, his training as a Marine had taught him the skills necessary to survive such an ordeal. She simply could not conceive of him being gone.

In time, Sally's disbelief turned to sorrow—and sorrow turned to anger. She felt that the site of the collapsed buildings was not being properly respected and that the "cleanup" was being rushed. After all, her son's remains—and those of countless others—were still in the rubble. And she was uncomfortable with the limited flow of information to families of civilian victims. She searched for meaning in a world that no longer made sense.

As Sally grappled with the reality of her loss—and as the country galvanized in anger—she discovered a newspaper article that roused her sensibilities. The Twin Towers, wrote two safety specialists, need not have collapsed, and an investigation into the cause should be conducted. Sally immediately contacted one of the writers, Glenn Corbett, an adjunct professor at John Jay College, and as they spoke, she began to grasp purpose in the wake of tragedy.

With single-minded determination, Sally embarked on a journey to uncover the causes of the Twin Towers' collapse. "Those buildings 'murdered my son,'" she tearfully told a reporter.[3] She blazed with curiosity and anger. How could buildings, engineered to survive the shock of a jetliner collision, simply crumple to the ground? Shouldn't the hulking steel frames of the structures been able to withstand the heat of a devastating fire? From the questions grew resolve.

A former nursing-home administrator with only minor participation in civic activities, Sally formed the Skyscraper Safety Campaign in December 2001 in memory of her son and immediately found herself at the forefront of the 9/11 organizational and investigative effort. "She's been going full throttle ever since, telephoning, writing letters, pressing Congress to investigate, assembling Corbett and others as the group's technical advisory panel, staying in contact with the other family members and staying on top of [the National Institute of Standards and Technology]," wrote the *Washington Post*. "She's become a walking compendium of World Trade Center trusses, beams, floor plans, fireproofing, emergency doors, evacuation plans, and codes on building and fire safety."[4]

Sally's campaign was never intended to be partisan or political, and it was focused more on fact-finding than finger-pointing. "Her fear, though," wrote the *Post*, "is that human error in the twin towers design or construction may have compounded the tragedy by leaving the buildings vulnerable to collapse. Christian's legacy, she says, 'has to be reform.'"[5]

Among the primary objectives of the Skyscraper Safety Campaign was the creation of a comprehensive independent federal investigative panel to study the construction of the World Trade Center towers and the integrity of the materials used in that construction. The campaign ultimately sought the improvement of codes, implementation of better-quality design practices, and enhancement of emergency response procedures and equipment.

On November 19, 2003, Sally Regenhard testified before the 9/11 Independent Commission in Washington, DC. She outlined the goals of her organization and lauded the swift reforms and safety measures that had already been instituted by the New York City Department of Buildings and its World Trade Center Building Code Task Force after 9/11, such as restrictions on truss construction in high-rise buildings and hardening of staircase and elevator cores. "As the Skyscraper Safety Campaign praises such actions, however, we are greatly dismayed that there is still, unbelievably, widespread resistance to national reform." She implored the leadership of the commission to become a catalyst for much-needed building-code enhancements throughout the country: "Generally, we are crawling, instead of running towards change," she said.[6]

Through Sally's dogged efforts and a mandate from the National Construction Safety Team Act, the National Institute of Standards and Technology (NIST) undertook "a technical probe" of the Twin Towers collapse in 2002. "At the usually cloistered [NIST], Regenhard's a familiar name," wrote the *Post*. "She speaks with [its] officials often."[7]

After three years of investigation, NIST issued a report revealing a sequence of failures that culminated in the collapse of the towers. Arising, of course, from the impact damage inflicted by the airplanes, the subsequent progression of flames, driven by jet fuel, and the displacement of fireproofing materials contributed to the sagging of floor trusses, from the intense heat of the rapidly escalating fire. As a result, the exterior columns on one side of each tower were gradually drawn inward until they reached a critical point of buckling. This, in turn, precipitated a cascade of instability, culminating in the swift descent of the upper segments onto the floors below.[8]

Along with these findings, the institute listed thirty building-code and fire-safety recommendations for state and local regulatory agencies. The proposals focused on updating fire-safety standards, strengthening retaining walls, promoting fire-resistant materials, and avoiding central stairwell cores. Though these proposals did not carry the binding weight of law, the institute implored state and local authorities to adopt the measures into their existing building-code policies.[9]

In a prepared statement on April 6, 2005, US Senator Hillary Rodham Clinton announced, "The Commerce Department's National Institute of Standards and Technology (NIST) presented the nation with a crucial document. By providing a preliminary report of how the World Trade Center . . . towers collapsed on Sept. 11, 2001, NIST took a significant step towards ensuring that we prevent future disasters and protect more lives if a disaster does occur."[10]

Though Clinton commended the Skyscraper Safety Campaign for its "tremendous leadership and . . . inspiration" on the subject of building safety, not all agreed with Senator Clinton's rosy assessment of NIST's work. Glenn Corbett, one of the Skyscraper Safety Campaign advisers, was decidedly disappointed with the report's lack of the specifics needed for codification. "You can't take this to a code group and say here. . . . Code groups deal typically in very specific numbers and sizes," he said. "It's going to be years before we get anything meaningful."[11]

The Citicorp Tower was engineered and constructed at a time when its designers could barely conceive of the possibility of terrorist attackers using fuel-laden commercial jetliners for the purpose of destroying metropolitan skyscrapers. Though the structure was designed to withstand external impacts, engineering for "worst-case scenarios" centered more on wind loads and seismic activity than on airplane collisions and prolonged burn.[12] The use of innovative architectural concepts and ultra-light-weight construction materials such as those employed by Stubbins and LeMessurier in the Citicorp Center should, nevertheless, have prompted a deeper analysis of the structural integrity of the building. As failures of human oversight may have contributed to the fall of the Twin Towers in 2001, could it also have played a role in the vulnerabilities of the Citicorp Building as perceived by its lead engineer in 1978?

The building codes of the day were, no doubt, antiquated, ambiguous, and largely inadequate, but most architects and structural engineers correctly viewed them as baseline requirements often exceeded by custom and practice.[13] The architect and educator Eugene Kremer wrote in 2002, "Like many other laws and regulations safeguarding public safety, building codes specify minimum standards and they do not necessarily reflect the state of the art or the prevailing standard of care."[14]

Kremer acknowledged that the governing New York City Building Code in the mid-1970s could fairly be interpreted as not requiring analysis of quartering winds, but he stated that many other high-rise buildings in New York and around the world at the time were still designed to withstand them.[15] Quoting the celebrated engineer Matthys Levy, author of *Why Buildings Fall Down*, Kremer wrote, "From the code point of view, it is implicit that wind from any direction should be considered, even if not stated explicitly."[16] Leslie Robertson, one of the engineers of the Twin Towers, agreed. "The New York City Code and nearly all other building codes, as well as good engineering practice, required the consideration of diagonal winds," he wrote regarding the Citicorp Building bracing system.[17] Indeed, LeMessurier himself, while serving on the Boston Code Committee in the 1960s, had written a provision requiring the consideration of forty-five-degree winds in the design of square or rectangular buildings. "This provision routinely applied to Citicorp would have automatically generated the 40% increase in diagonal force," he later admitted.[18]

Though LeMessurier maintained that he and his associates did consider quartering winds in evaluating the midface placement of the four principal columns, he determined that such winds "did not govern the design and need not be further considered."[19] Furthermore, he continued to insist that neither New York code nor industry practice required consideration of nonperpendicular winds on any component of the structure, including bracing or bolt design. "This practice," he said, "is almost universal in design offices for square and rectangular buildings."[20] Both of these factors, LeMessurier ultimately conceded, created a "design tradition and mental set" that led his engineers to conclude that *no* element of the building was vulnerable to quartering winds.[21] "The sensitivity of the wind bracing system to non-perpendicular winds," he thus wrote, "was never perceived by any member of the final design team."[22]

While most of the engineering designs and computations for the Citicorp Tower were produced by LeMessurier's New York team, it was clearly his personal obligation and that of the Office of James Ruderman as engineers on the project to ensure that every reasonable safety measure was employed in the tower's design. And while LeMessurier insisted that informal checks between the two offices were conducted and that all drawings were stamped by Murray Shapiro, Ruderman's principal, he did acknowledge that "there was no formal procedure for checking work either within or between offices."[23] "My mistake was in designing a structure that was innovative, and I didn't check and dog people carrying it through carefully enough," he later said.[24]

The failure of LeMessurier and his engineers to appreciate the effects of quartering winds on the Citicorp bracing system and connecting bolts may, in fact, have been only part of the problem. According to the author Siobhan Roberts, in her book *Wind Wizard*, the wind analysis report issued by Alan Davenport and his team at the Boundary Layer Wind Tunnel Laboratory as part of the initial design phase of the building indicated the need for consideration of perpendicular winds as well as diagonal winds, applied either individually or simultaneously. "Bill had not taken into account the combined action of wind-induced forces acting in both principal directions of the building," recalled Davenport's associate Nicholas Isyumov. "The need to consider this joint action was clearly reported in our wind tunnel study. His structural system was equally sensitive to both sets of these forces and was therefore overloaded."[25]

And yet modern inquiry perhaps challenges these conclusions. In 2020, the research engineer Dat Duthinh conducted extensive wind-tunnel analysis of the Citicorp Tower using state-of-the-art computer technology developed at NIST to simulate real-world, long-duration, dynamic wind loads and concluded, notwithstanding LeMessurier's concerns, that face winds placed *more* structural demand on the wind-bracing system than quartering winds did. "The results surprised us," wrote Duthinh. "The analysis . . . shows that corner winds are less demanding than face winds."[26] Though Duthinh's conclusions have been questioned by engineers familiar with the design and specific vulnerabilities of the Citicorp Tower, the controversy continues.

Robert McNamara, LeMessurier's former associate, questioned his boss's response to the issue of quartering winds. "Bill is one of the finest engineers I've ever worked for," said McNamara in 2023. "I travelled the world with him and I knew him quite well and I don't know why he went that route."[27]

◢

Just ten days after the 9/11 attacks, a professor of civil and environmental engineering at MIT, Oral Buyukozturk, asked the fundamental question: "How safe are our skyscrapers?" The question was just as relevant in the mid-1970s as it was following the collapse of the Twin Towers. "The key," wrote Buyukozturk, "is built-in redundancy in all system components of a complex system such as a skyscraper."[28]

While the concept after 9/11 applied more relevantly to fireproofing, excavation plans, and structural systems designed for slower rates of failure, redundancy as viewed in the context of the Citicorp Tower can be thought of on multiple levels. First, and most obvious, LeMessurier acknowledged that the failure of *one* bolted connection of the bracing system could lead to the progressive failure of the entire structure. Though the originally designed welded connections would have been more than adequate to withstand quartering winds, once the design was changed to bolts, the issue of redundancy became a much more acute concern. The question of how the lapse occurred also involves consideration of redundancy on the human level.

LeMessurier steadfastly denied knowledge of the design change to bolted connections, but whether he knew or not misses the point that he *should* have known. And if he did know, much greater oversight in the engineering of those bolts should have been provided. Diane Hartley noted that the shop calculations provided to her by the engineers were initialed by only one member of the firm and, evidently, not crosschecked by others. Redundancy of *effort* should have been employed by the engineering team in developing, checking, and rechecking wind-load concepts and calculations. Joel Weinstein was a talented and perfectly competent young project engineer, but he and his colleagues had been tasked with implementing a building design concept generated by others in the firm's Cambridge office. LeMessurier conceded that there was no formal inter- or intraoffice procedure for verifying the accuracy of engineering work. A "design tradition and mental set," as he called it, are clearly an insufficient justification for overlooking a principle with potentially catastrophic ramifications, especially when that tradition and mental set are not universally shared in the industry. LeMessurier's failure to conduct a more comprehensive analysis of quartering winds in the design of the Citicorp Tower or its splice bolts can, thus, be attributed, in part, to a failure of oversight and the absence of redundancy in work procedures.

◢

Following 9/11, there emerged a perception that the era of the modern skyscraper had drawn to an abrupt close. With the United States' high-rise buildings viewed as the new primary target of international terrorists, many people believed that these structures had become dangerous ground. The sight of the World Trade Center towers crumbling due to acts of unspeakable and intentional violence led some people, including those in the construction sector, to speculate that our willingness to inhabit, work in, and gather within such visible and conspicuous structures might vanish.[29]

The reality, however, was starkly and surprisingly different.

With surging urban populations and property values, the economic pressure to push the limits of building height became inevitable. According to the Council on Tall Buildings and Urban Habitat, a nonprofit association focused on the built environment and the growth of "vertical urbanism," 84 percent of all buildings two hundred meters or taller currently in service were erected *since* 2001, including eighty-six of the world's one hundred tallest buildings and eight of the ten tallest buildings in the United States.[30] In fact, in the twenty years following 9/11, the average height of the world's one hundred tallest buildings has *increased* by 141 percent. The Twin Towers of New York's World Trade Center would only be, if standing in 2021, the seventh and eighth tallest buildings in the country.[31]

While the skyscraper is decidedly not dead, it has been forced to adapt. The number of high-rise buildings employing all-steel construction has dropped to 9 percent since 2001, with a substantial increase in the use of concrete and composite materials.[32] Climate change, economics, and worldwide health risks have affected the design and construction of the modern urban landscape. During the COVID-19 pandemic, examination of the "future-proofing and sustainable development of our cities" pervaded the work and attitude of architects and urban designers and prompted land developers to adapt to an uncertain world.[33] "The introduction of new codes and regulations, the call for a holistic approach to sustainability and the increased incorporation of technology—will play significant roles in how our cities develop in the near future," wrote one architect.[34]

Development, however, does not occur in a vacuum. As Walter Wriston resolved with the Citicorp Center, vertical growth should never forsake the community below. Fiscal equity in land development—the embrace of fair economic treatment of urban neighborhoods—should play a fundamental role in good city planning. "If we're going to redevelop downtown," wrote the Pulitzer Prize–winning architecture critic Blair Kamin, "let's try to redevelop it for a broad segment of people, not just the rich."[35]

In many respects, the design concept of the Citicorp Center was years ahead of its time. Energy efficient, architecturally innovative, and socially enlightened in the provision of private-public spaces, the complex would serve as a shining example of forward thinking in urban design.

In June 1978, however, William LeMessurier wondered if that design contained a subtle but fatal flaw.

8

ONE IN SIXTEEN

JULY 25–27, 1978

In 1973, because of the lagging economy, LeMessurier had reluctantly negotiated a merger of his company with a mechanical and electrical engineering firm called Sippican Consultants International. As a result, he had lost some of his managerial autonomy. Now, he was hesitant to tell his corporate partners about the potential problem with the Citicorp Tower, fearing they would try to keep him quiet. "I know in my heart of hearts if I'd gone to them they'd have told me to keep my mouth shut," he later said.[1]

One trusted and respected associate, Al Romaneski, a former colonel in the US Army and LeMessurier's hand-picked "second in command," however, pledged his unwavering support when LeMessurier confided in him. A West Point graduate and a man of impeccable detail and moral standing, Romaneski would prove to be a loyal and indispensable ally, even as he served as a Sippican officer. As LeMessurier told it, "I said Al, this is what I have to do. He said okay, I'll follow you up the hill. . . . That was very warming to have a man of that caliber who understood the corporate relations we had."[2]

LeMessurier feared that he may have oversimplified the problem with the Citicorp Tower by what he referred to as his "little geometrical game." While his calculations appeared to be theoretically sound, he recognized that they may not be universally applicable. He was aware that wind loads on tall buildings vary in form and intensity, and a straightforward calculation for one scenario may not translate seamlessly to all situations.[3] What troubled him most were the anomalies, particularly the unique oscillations induced by unpredictable and sustained turbulence interacting with the tower. "A building is like a tuning fork," he said. "What's really happening is that the building is vibrating this way and that, it's when the vibrations gang up [in a particular] direction that you get those forces peaking in certain diagonals."[4] Mathematical probabilities are one thing, he thought, but real-world conditions are quite another.

LeMessurier recognized that wind could exert diverse influences on a structure, necessitating a customized analysis of dynamic forces for each specific building and condition. Once he determined that the tower's bracing system contained an anomalous susceptibility, he required quantifiable data capable of statistically predicting the likelihood of critical-level wind stresses on the chevron connecting bolts.

On July 25, 1978, LeMessurier telephoned Alan Davenport and briefly explained his discovery and concerns with the Citicorp Tower. He asked that the Wind Tunnel Lab team gather all of its files, test results, and magnetic storage media on the building, and he requested an immediate meeting. "We knew it was an urgent problem," Davenport later recalled.[5]

The following day, LeMessurier flew to London, Ontario, hoping to abate his fears. He nervously asked Davenport and his associate Nick Isyumov to reanalyze their wind-tunnel data in light of the impact of quartering winds on the connecting bolts of the Citicorp Tower bracing system. Assuming that the structure did not possess the same level of strength and rigidity as previously thought, the question was what degree of wind stress could it now withstand. "You have to tell me the truth," he told them. "Don't be easy if it doesn't come out the right way."[6]

LeMessurier took some comfort in the protection provided by the TMD. The splice bolts had been designed, inadequately as they were, under the assumption that the TMD was *not* operational. In point of fact, LeMessurier knew that the machine would not activate if the power in the building was out because the oil pump would not operate. He thus requested the wind technicians to generate two sets of calculations: one with damping and one without.

After two days of rigorous testing and analysis, the news was not good. Though LeMessurier's computation of a 40 percent increased load caused by quartering winds was technically accurate, it was, according to Davenport, only a baseline approximation. As LeMessurier had suspected, the actual real-world stresses could be even higher depending on the particular weather conditions in play. And when the erroneous gravity-compression figure utilized by his New York engineers during calculation of the bolt strength was factored in, the results could be exponentially worse.

LeMessurier's mind was swimming. Having gathered the refined wind-tunnel data for each structural member of the Citicorp Tower wind-bracing system from Davenport, he now needed to meticulously—and privately—analyze it.

"I forced myself to keep cool until I had all the facts," he later said.[7]

JULY 28–30, 1978

As a child in Michigan, LeMessurier often visited Higgins Lake in Roscommon County with his family. It was (and still is) a pristine and enchanting vacation spot in the heart of the state. Later, moving to the Northeast and raising his family, he sought a similar destination of his own. In 1964, he purchased a summer getaway off the northern shores of Lake Sebago, Maine—a fourteen-acre island named Doctor Island, with a cottage in need of repair. A few years later, he rebuilt the structure, imbuing it with his own distinctive design, featuring a cross-butterfly roof and magnificent views from nearly every angle. Later, he purchased a small parcel on the mainland adjacent to the island to park his car and speedboat, which he used to shuttle to and from the house. Doctor Island would become LeMessurier's personal sanctuary, his refuge from the rigors of life.

Upon LeMessurier's return to Cambridge from his meeting in Canada, he briefed Al Romaneski and then telephoned his wife, Dorothy, at their cottage in Maine. He confided in her about the potential vulnerability of the Citicorp Tower and his unease about the safety of the building. Dorothy keenly understood the pressures that her husband's career often brought, and she worried about the toll the situation was now taking on him. He had decided to spend the weekend at their lakeside home reviewing the data and recalculating the quartering wind loads on the building. "I felt I had to go into a retreat and think it all through," he said.[8]

Through the entire drive north, LeMessurier ran numbers and scenarios through his mind and fought the incessant rise of anxiety.[9] For much of the weekend, alone in his lakeside office, he worked through Alan Davenport's wind-tunnel data. "What I wanted to know," he later said, "was when was this building going to fall down."[10]

LeMessurier performed a detailed, floor-by-floor, brace-by-brace analysis of the structure and calculated the actual wind forces relative to the ultimate strength of the various diagonal splices. He had learned that isolated wind gusts alone would not generally cause the structural failure of a tall building. They may knock out panes of glass or blow off some exterior sheathing, but they would not, in all likelihood, destroy it. It was the sustained winds that most troubled him: the winds that cumulatively envelope a structure and, over a period of time, cause it to oscillate and sway beyond its designed limits.

LeMessurier's focus quickly centered on the thirtieth story, which contained, in his estimation, the most vulnerable bolted joints. Failure of those joints, he realized, would result in the progressive collapse of the entire tower. He then nervously assessed the likelihood of a storm in New York City strong enough to cause the thirtieth-floor bracing bolts to stress to the point of fracture. Through careful examination of weather statistics provided by Davenport, he determined that a fifty-year storm could cause a catastrophic failure event when the stabilizing effect of the TMD was considered: a one-in-fifty chance in any given year. Removing the electric-powered damper from the equation—as a destructive storm could absolutely do with a loss of power—the odds of failure increased to a staggering one in *sixteen*. "That was very low, awesomely low," he recalled.[11]

Staring blankly out the window of his island retreat, and grimly aware that the Atlantic hurricane season was fast approaching, LeMessurier considered the prospect of professional disgrace, financial ruin, and the horror of a building collapse in the heart of New York City.

9

"A CATALOGUE OF FAILURES"

LeMessurier would inevitably ask himself how he had even considered a bracing system design without fully assessing its potential vulnerabilities. The question would haunt him for decades. Schooled in the history of architectural blunders and structural failures, he understood, perhaps better than anyone, the potential risks of miscalculation and poor business practice. Though he consistently argued that neither the New York City Building Code nor engineering custom and procedure mandated consideration of nonperpendicular winds in the design of tall buildings, privately he believed the contention would not absolve him of blame for his oversights. "[I] myself [have] been fully aware of the significance of winds from all directions for many years," he would confess.[1]

In 2013, Henry Petroski, a structural engineer and educator specializing in failure analysis, wrote, "The engineer in possession of a catalog of failures, however old, is an engineer equipped with an anthology of horror stories and an arsenal of arguments that may be called to bear when a colleague suggests even a supposedly minor design change. The seemingly innocuous change . . . might appear to be an obvious improvement to the engineer making the suggestion, but to the engineer schooled in failure case studies it should be a red flag."[2]

LeMessurier bore the burden of the Citicorp Tower vulnerability with awareness of the disconcerting history of global structural failures and their catastrophic aftermaths. His actions and omissions in the design and evaluation of the Citicorp Center may, therefore, be best viewed in the context of that "catalog of failures."

◢

As early as AD 27, the calamitous effects of structural failure had already been chronicled. North of Rome in the city of Fidenae, an amphitheater constructed to showcase gladiator combat collapsed without warning, killing an estimated twenty-thousand people. The tragedy was graphically recounted by the Roman historian Tacitus:

> In the year of the consulship of Marcus Licinius and Lucius Calpurnius, the losses of a great war were matched by an unexpected disaster, no sooner begun than ended. One Atilius, of the freedman class, having undertaken to build an amphitheatre at Fidena for the exhibition of a show of gladiators, failed to lay a solid foundation and to frame the wooden superstructure with beams of sufficient strength; for he had neither an abundance of wealth, nor zeal for public popularity, but he had simply sought the work for sordid gain. Thither flocked all who loved such sights and who during the reign of Tiberius had been wholly debarred from such amusements; men and women of every age crowding to the place because it was near Rome. And so the calamity was all the more fatal. The building was densely crowded; then came a violent shock, as it fell inwards or spread outwards, precipitating and burying an immense multitude which was intently gazing on the show or standing round. Those who were crushed to death in the first moment of the accident had at least under such dreadful circumstances the advantage of escaping torture. More to be pitied were they who with limbs torn from them still retained life, while they recognised their wives and children by seeing them during the day and by hearing in the night their

screams and groans. Soon all the neighbours in their excitement at the report were bewailing brothers, kinsmen or parents. Even those whose friends or relatives were away from home for quite a different reason, still trembled for them, and as it was not yet known who had been destroyed by the crash, suspense made the alarm more widespread.[3]

The Fidenae amphitheater collapse, considered the deadliest structural failure in recorded history, was attributed, as per the sole historical account and subsequent commentary, to an avoidable engineering blunder driven primarily by inattention and greed. While modern building codes, government oversight, and physical science have come a long way to curtail and prevent such disasters, structural failures have unfortunately continued through history even to this day.

◢

The Tacoma Narrows Bridge disaster, studied by Diane Hartley and engineers, physicists, and mathematicians worldwide, would become a cornerstone in the research of wind effects on manmade structures. Opened in July 1940, "Galloping Gertie," as it was inauspiciously nicknamed because of its dramatic undulations during wind conditions, stretched across Puget Sound in western Washington State. At the time, it held the distinction as the third-longest suspension bridge in the world.[4] In the late morning of November 7, 1940, just four months after its completion, the bridge dramatically collapsed in a wind storm with, remarkably, no loss of human life. "I saw the Narrows Bridge die today," reported Leonard Coatsworth, news editor of the *Tacoma News Tribune*, "and only by the grace of God, escaped dying with it."[5]

F. B. "Bert" Farquharson, a professor of engineering at the University of Washington who, ironically, was photographing the bridge at the time of its failure as part of a commissioned study of its movement, wrote,

I was the only person on the Narrows Bridge when it collapsed. . . . When I arrived at about a quarter to ten o'clock, the bridge was moving in the familiar rippling motion we were studying and seeking to correct. . . . About a half hour later, it started a lateral twisting motion, in addition to the vertical wave. It had never done that before. . . . At least six lamp posts were snapped off while I watched. A few minutes later, I saw a side girder bulge out. But, though the bridge was bucking up at an angle of 45 degrees, I thought she would be able to fight it out. But, that wasn't to be. . . . I saw the suspenders (vertical cables) snap off and a whole section caved in. The bridge dropped from under me.

Clark Eldridge, supervising engineer and chief advocate for the construction of the Tacoma Narrows Bridge, wrote in his 1986 memoirs, "I go over the Tacoma bridge frequently and always with an ache in my heart. It was my bridge."[6]

Tacoma Narrows has become a classic case study in the field of civil engineering and structural design. The failure was originally attributed to "random action of turbulent wind," but an actual particularized cause has been elusive. It is generally agreed that excessive flexibility of the bridge's components made it susceptible to a torsional or twisting motion. In the moments prior to its collapse, the bridge was exposed to rising winds and alternating vortices or swirling patterns of air that began lifting and twisting the bridge deck. At about 10 a.m., a centrally located cable band slipped and separated into two segments of unequal length, resulting in "torsional flutter"—a self-intensifying rhythmic or harmonic oscillation. As the convulsion gradually increased, the structural capacity of the bridge was exceeded, resulting in its total failure.[7]

The Tacoma Narrows Bridge collapse was jarring to both everyday citizens and structural designers. At the time, the study of aerodynamic forces on buildings and bridges was in its infancy, and meaningful preconstruction wind analysis was rare. A three-member panel of engineers appointed by the Federal Works Administration (FWA) to investigate the Tacoma failure recommended among its findings that wind-tunnel testing of structural designs should become part of industry practice.

Bridge and building designers would heed the FWA's advice. The collapse would lead to significant advancements in the understanding of aerodynamics and structural engineering, and it played a critical role in improving the safety and stability of subsequent suspension bridges. "Regrettable as the Tacoma Narrows Bridge failure and other recent experiences are," said Othmar Ammann, the renowned designer of New York's George Washington Bridge and a member of the FWA panel, "they have given us invaluable information and have brought us closer to the safe and economical design of suspension bridges against wind action."[8]

As T. R. Witcher, contributing editor of *Civil Engineering*, wrote in 2017, "there's obviously no good time for a bridge to fail, but the Silver Bridge's collapse, on Friday, December 15, 1967, at 4:58 PM, was about the worst moment possible."[9]

The Point Pleasant Bridge, known more commonly as Silver Bridge because of its distinctive aluminum-hued paint finish, was dedicated on Memorial Day 1928. Spanning the Ohio River and connecting Point Pleasant, West Virginia, and Gallipolis, Ohio, the bridge would bring community pride and economic growth to the towns and villages that it served.

Though the bridge was initially designed with conventional wire suspension cables, contractors were encouraged when construction bids were solicited to propose alternative bridge designs with lower costs, in exchange for an equal share of the savings. The winning bid, submitted by the American Bridge Company, featured a double-lane roadway deck suspended from two parallel chains composed of two-inch-thick steel links, each forty-five to fifty-five feet in length and coupled with steel pinions eleven inches in diameter.[10] Many of the Silver Bridge eyebar couplings unconventionally also functioned as an element of the bridge's stiffening system. With the employment of a unique process of heat-treating the steel elements, it was believed that the overall structure would be stronger, lighter, and less costly. Engineers of the day were convinced that the bridge, so constructed, was significantly overdesigned and, thus, virtually immune from failure.[11]

Through the years, state officials conducted sporadic (and sometimes deficient) inspections of the Silver Bridge. Unsurprisingly, those inspections revealed no significant structural issues, notwithstanding the fact that the weight of the typical automobile using the bridge had dramatically increased since its original design in the mid-1920s. Through the years, no concerning report pertaining to the bridge's structural integrity or stability had been documented.

The "worst moment possible" as described by Witcher came as the Silver Bridge was teeming with Friday-afternoon commuters and holiday shoppers. Witnesses heard what they described as a gun-shot sound, and then the bridge began to fold on itself. Charlene Wood, four months pregnant and heading home after a visit with her parents, had just driven on the first span of the structure when she felt a fierce shudder. She quickly rolled her car backward and watched in horror as the bridge crumbled in front of her. "It was like someone had lined up dominos in a row, and gave them a push," she recalled, "and they all came falling down and there was a great big splash of water. I could see car lights flashing as they were tumbling into the water. The car in front of me went in. Then there was silence."[12] Thirty-two vehicles fell into the raging river below, taking forty-six victims to their deaths. It would be described as "the most tragic highway bridge accident in U.S. history."[13]

From disaster, however, came action. President Lyndon Johnson quickly launched the President's Task Force on Bridge Safety to investigate the collapse and to evaluate the general state of the nation's bridges. The investigative team conducted a detailed examination of the Silver Bridge wreckage and quickly turned their focus to the steel eyebars composing the bridge's suspension system. Three years after the collapse, the task force, in conjunction with the National Transportation Safety Board (NTSB), released a comprehensive final report, which included a detailed chemical and mechanical analysis of the components of the bridge and a methodical chronology of events that ultimately led to its failure. Citing a small cleavage fracture in eyebar 330, caused by the combined effects of "stress corrosion and corrosion fatigue," the board determined that the insidious decomposition and ultimate failure of that eyebar led to the dislodging of the steel pin connecting it to the next link, thus severing the north suspension chain. "With the . . . chain thus broken . . . the structure's design made total collapse . . . inevitable."[14]

While the absence of engineered redundancy in the suspension chain system of the Silver Bridge undeniably played a significant role in its failure (the loss of a single link inevitably resulted in the collapse of the entire structure), the NTSB appeared to absolve the design team of any accountability. As contributing causes of the bridge collapse, the report observed,

1. In 1927, when the bridge was designed, the phenomena of stress corrosion and corrosion fatigue were not known to occur in the classes of bridge material used under conditions of exposure normally encountered in rural areas.
2. The location of the flaw was inaccessible to visual inspection.
3. The flaw could not have been detected by any inspection method known in the state of the art today [1970] without disassembly of the eyebar joint.[15]

The tragedy of the Silver Bridge collapse would nonetheless lead to much-needed reform. The NTSB's final report recommended that the secretary of transportation establish initiatives to study the load-bearing capacity and life span of the nation's bridges, as well as the vulnerability of bridge construction materials to long-term corrosive forces. It also promoted the establishment of national bridge safety standards, mandatory safety inspections, and the allocation of federal funds to facilitate all required repairs.[16]

In 1968, as a direct consequence of the Silver Bridge disaster, Congress authorized the National Bridge Inspection Standards Program—the country's first nationally coordinated bridge safety initiative.

Petroski wrote, "Failure is more often like a bad dream. It can progress silently behind our tightly closed eyelids, disturbing no one but ourselves, until suddenly it reaches an intolerable state, at which time something snaps and we are startled awake screaming."[17] The Tacoma Narrows and Silver Bridge collapses graphically demonstrate the evolving nature of structural science. The imperative is to learn from these failures and to find solutions to their causes.

Structural engineers are not omnipotent. They make mistakes and are necessarily limited by the state of the science they rely on and the base of knowledge they possess. The Tacoma Narrows collapse introduced the study of aerodynamic forces on bridges, allowing safer and better-designed future structures, while the Silver Bridge disaster ushered in an era of oversight and meaningful inspection. But both graphically demonstrate the ever-present danger of the unknowable risk. They demand that science and public policy become partners in the quest for better structural design.

As innovation continually pushes the envelope of engineering, however, latent vulnerabilities continue to be discovered—and often too late. The use of modern building materials and techniques, competing with the unending quest for cost savings in design, has resulted in elegant and sustainable structures but often with increased risks. And in some cases, government oversight has not kept up with design innovation.

The Northridge, California, earthquake of 1994 and the Kobe, Japan, earthquake the following year both starkly revealed a slow and insidious erosion of design standards that had taken place throughout much of the twentieth century, despite a technological explosion in the field of structural science. In both of those locations, engineers had observed an unusually high number of fractures in steel-frame buildings caused by the seismic activity, particularly in spaces near the welded connections between girders and supports. Subsequent investigations concluded that a gradual decline of quality assurance for steel tolerance, welding technique, and construction standards had contributed to the vulnerability of these structures. "It was a classic case of devolution from a robust design that was incrementally eroded in the name of economy and efficiency," wrote Petroski.[18]

There is ample evidence that structures can be designed with due care and in full compliance with code and practice and still ultimately fail. But what of blatant negligence in the name of expedience or profit?

For over a century, there had not been a deadlier building failure in the United States than the tragic collapse of the Pemberton Mill in 1860, which claimed the lives of an estimated 88 to 145 workers in Lawrence, Massachusetts. It was, according to one commentator, "the most heart-rending calamity of the age."[19]

The mill had been constructed only seven years before its failure, and the builders had utilized inferior cast-iron columns and substandard brick for walls. At 4:45 p.m. on the cold winter afternoon of January 10, multiple floors, overloaded with ton after ton of textile machinery, buckled and crumbled to the ground. John Ward, a worker in the mill, informed the members of the inquest that followed,

> Suddenly, I heard a noise—it sounded like a loud, thundering crash over my head, and looking up, I saw the shafting coming down upon us, all over the room. I could not account for it, and was terrified. I stood nailed to the spot, and did not seem to have power to move, although I knew the building was coming on me. Then I heard the overseer shout, and I tried to jump out of the rubbish, but something struck me, and I was thrown senseless. I did not remain so long, but when I came to, I found myself buried in the rubbish, and did not expect to get out alive. I was all covered over with blood, from wounds on my face. I finally crawled up, and got to the top, and found a lot of ruins hanging over me, which like to have

taken my life. But I succeeded in getting out. I passed by a dead girl on my way, and two other mangled bodies before I got out. When I was first knocked down, I fell beneath a large grinding stone, which was too heavy to give way to the weight above, and this saved my life. When I fell under there, I saw the walls over me all falling, and the floor giving way all around me.[20]

Earliest public blame for the disaster came from none other than local clergy. The Reverend E. M. Tappan of the Freewill Baptist Church pulled no punches in his Sunday-morning sermon. "From the testimony before the coroner, now holding an inquest, it appears that the Pemberton Mill was not substantially built," he claimed.

The walls were too thin—much thinner and weaker than those of other mills; the pillars were not strong enough. From these causes, undoubtedly, it fell. . . . Why did it fall? We are told by some that this disaster is a judgment of God upon this city for its wickedness. Well, if I should leap from the bridge into the river, and lose my life, that might just as well be called a judgment of God, because he did not suspend the law of gravitation in my behalf. This mill was built in violation of physical laws, and it fell just as such buildings always must fall. But why was it thus built? Because it would *cost* more to build it substantially. The *inordinate love of money* is the cause of this sad disaster. Those persons did not lose their lives because they were worse than others; that mill did not fall because its owners were worse men than the owners of the other mills in our city; our city does not suffer this calamity because it is worse than all other cities; but because, to save expense, the mill was not properly built. And if there is not a change, other and similar accidents will still occur. There is too much haste to be rich; hardly any one can be trusted. . . . Now, unless there is a change in this respect, we may expect scenes like those just witnessed. . . . Our State should appoint Building Commissioners to look to these public buildings.[21]

Reverend Tappan's prescient comments are as relevant today as they were in 1860. While the science of structural engineering has unquestionably made significant strides since the Pemberton Mill disaster, work remains. The evolution of new and exciting building formats and designs must continue with unfailing government oversight and without erosion of sound engineering practice.

▲

Perhaps the most glaring modern example of negligent structural design, and the one that would later demonstrate William LeMessurier's gravest fears in 1978, is that of the Hyatt Regency Hotel disaster in Kansas City, Missouri.

On the evening of July 17, 1981, beneath the gracefully suspended "skywalks" of the city's chic new hotel, the brass melodies of big-band swing filled the air. The picturesque atrium was bustling with partygoers who had gathered for that night's "tea dance," delighting in the music of the Steve Miller Orchestra and seeking respite from the sweltering midwestern summer night. A dance contest was under way, with participants swaying gracefully to the rhythm of Duke Ellington's "Satin Doll," and curious hotel guests had congregated on the three elevated walkways that bridged the atrium watching the festivities below. Just after seven o'clock, a woman sitting in a stairway next to the lobby turned to her husband and said, "Isn't this a nice place to spend the evening?"[22]

Moments later, witnesses were startled by a sharp report that cut through the music like a thunderclap. In an instant, the fourth-floor walkway spanning the atrium buckled at its midpoint, dropped slightly, and then tumbled down onto the lower walkway, causing both structures to plummet directly into the gathering below. "The place was packed," said one observer. "Suddenly there were people and water flying everywhere. . . . It hit the second one and there must have been 75 or 100 people on that. . . . And then that fell and there must been 100 to 150 people below that. It all looked like a human sandwich."[23]

The plummeting morass of steel, concrete, and glass weighed an estimated 142,000 pounds. In total, 114 people lost their lives in the collapse, and 186 were physically injured.[24] But the emotional wounds were incalculable. In one way or another, roughly half of Kansas City's residents were believed to have been impacted either directly or indirectly by the Hyatt Regency collapse.[25]

The ensuing investigation conducted by the National Bureau of Standards found that the system of supports for the suspended skywalks was woefully underdesigned. The original plans created by the structural engineers, Jack D. Gillum and Daniel Duncan, and approved by the Kansas City Codes Administration Office called for a single set of six steel hanger rods fastened to the ceiling and connected to the fourth- and second-floor box beams below. "Under this arrangement," the bureau wrote, "each box beam would separately transfer its load directly into the hanger rods."[26] The design of these connections required precise engineering calculations to determine their adequacy to carry the loads for which they were designed.

During construction, however, the steel fabricator submitted revised plans changing this configuration to include two sets of shorter hanger rods, one from the ceiling to the fourth-floor beams and another from the second-floor beams to the fourth-floor beams—thereby shifting the full load of *both* walkways to the ceiling hanger rods. In effect, the new design doubled the stress on the upper-level suspension apparatus. The report noted that the revised shop drawings were stamped by the general contractor, the architect, and the structural engineers.

In all probability, the change in design did not cause the disaster; it only hastened it. The hanger-rod connections to the box beams were designed without redundancy, meaning that if any connection within a single walkway were to fail, it would trigger a cascading failure, leading to the collapse of the entire walkway. The bureau concluded that *neither* support design complied with the minimum standards of the Kansas City Building Code. But, after installation of the revised hanger-rod configuration, "the ultimate capacity of the walkways was so significantly reduced that, from the day of construction, they had only minimal capacity to resist their own weight and had virtually no capacity to resist additional loads imposed by people."[27]

In the wake of the disaster, disciplinary proceedings against the engineers Gillum and Duncan were initiated. Both denied reviewing the revised plans, though each admitted knowledge of and acquiescence to the change. Though the men were ultimately acquitted of any criminal charges, the Missouri Board for Architects, Professional Engineers and Land Surveyors found them guilty of gross negligence and misconduct in the performance of their professional duties and revoked their licenses to practice. In affirming the board's actions, the Missouri Court of Appeals noted that the Kansas City Building Code was conceived to provide a minimum level of safety for the city's buildings. The engineers' "gross failure" to comply with the code, wrote the court, was tantamount to "conscious indifference to duty." And in what should be construed as a broad admonition to the engineering profession as a whole, the court concluded, "The level of care required of a professional engineer is directly proportional to the potential for harm arising from his design."[28] The Hyatt Regency disaster presents one of the most frightening cautionary tales for structural engineers and building designers worldwide. It was and still remains, according one commentator, the "quintessential 'teachable moment.'"[29]

As catastrophic structural failures continue to occur with alarming frequency, the obvious questions remain: Why do they occur, who is responsible, and how do we prevent them?

The actual causes of individual failures vary, of course, depending on the structures involved and the specific circumstances in play. Among the general causes of insidious and progressive structural problems are faulty workmanship, defective materials, improperly engineered specifications, and inherent design flaws. Clearly, some failures through history have been the result of unknowable design defects that were beyond the state of structural science. Such was the case with the Tacoma Narrows Bridge and, to some extent, the Silver Bridge. Designers must perceive the danger in order to prevent it. A methodical regimen of inspection of older structures is imperative to prevent the undetectable, insidious defect from ultimately becoming the source of catastrophic failure.

As was tragically demonstrated by the 2021 collapse of the Champlain Towers South Condominium in Surfside, Florida, where ninety-eight lives were lost, some of the lessons of the past have yet to be heeded. Similar to the Silver Bridge collapse, the Surfside tragedy can be attributed, at least in part, to the gradual deterioration of structural elements over time, as well as the failure on the part of inspectors and engineers to fully perceive the extent of the danger posed.

Other failures, such as Pemberton Mill and the Kansas City Hyatt Regency, show the devastating consequences of engineering misconduct and gross negligence—where the simple exercise of due care would and should have averted disaster.

The code of ethics of the National Society of Professional Engineers provides among its fundamental canons that engineers shall "hold paramount the safety, health, and welfare of the public" and shall "perform services only in areas of their competence."[30] Current law would require no less.

The tragic lesson of the Hyatt Regency, disastrously learned by the building's designers, is that inattention to detail or breakdown of process and procedure in matters of structural design can lead to fatal outcomes. It underscores the imperative of meticulously calculating and adhering to factors of safety mandated by local and state codes so that structural integrity can exceed even the harshest foreseeable stresses. Simple office procedures that foster open and honest communication can mean the difference between discovering inaccuracies and overlooking them. As Petroski wrote, "Engineers . . . are not superhuman. They make mistakes in their assumptions, in their calculations, in their conclusions. That they make mistakes is forgivable; that they catch them is imperative. Thus it is the essence of modern engineering not only to be able to check one's work, but also to have one's work checked and to be able to check the work of others. In order for this to be done, the work must follow certain conventions, conform to certain standards, and be an understandable piece of technical communication."[31]

In an ever-evolving urban landscape marked by cutting-edge structural designs and economic pressures that dictate engineering and construction practices, it is vital that designers continually reassess conventional techniques and approaches. Energy efficiency, sustainability, and climate resiliency are affecting building methods and materials. Engineering approaches that were once utilized with confidence may no longer be appropriate—or safe—in light of current architectural trends. For example, Boston University's nineteen-story Center for Computing & Data Sciences, which opened in January 2023, was curiously designed to resemble a heap of unevenly stacked books, with each floor asymmetrically skewed over the next in a series of dramatic cantilevers. The architectural and engineering challenges presented by this and other unique structures necessarily require consideration of specific risks and concerns beyond those of the conventional square or rectangular building. Traditional systems and practices suitable for older designs may need to be reevaluated in light of modern and evolving trends. Though basic design principles are axiomatic, the era of one-size-fits-all engineering has passed.

Structural failures become more likely when weakness, defects, or poor workmanship undermine their integrity. Natural disasters such as earthquakes or hurricanes can sometimes exert forces beyond designed resilience. The technology exists to overcome these dangers. The question, of course, is whether such measures are within economic realities. The integration of robust structural systems capable of withstanding extreme stresses has become an essential requirement for some contemporary skyscrapers. Buildings like 181 Fremont in San Francisco offer resistance to earthquakes through high-tech methods, while owners reap the benefits of increased property values and rental rates as a result. Beyond such approaches, basic practices such as quality control for construction materials, skilled and experienced workers, adherence to codes, and routine inspection and maintenance all combine to bring confidence and safety to owners and occupants of high-rise buildings.

As engineers strive to prevent structural failures, they must continually incorporate redundancy into their designs, ensuring that if a single component proves vulnerable, the integrity of the entire structure will not be jeopardized.[32] In some cases, compliance with the minimum standards set out in state and local building codes is not enough to satisfy the standard of care necessary for good engineering. Environmental factors, weather, climate, and seismic activity all play a role in urban design.

◢

If, as the Missouri Court of Appeals emphasized in upholding the license revocations of the Kansas City Hyatt Regency structural engineers, the standard of care expected of engineers and building designers rises in relation to the potential for harm, then it is evident that William LeMessurier likewise shouldered an increased level of responsibility for the adequate supervision of his New York engineering team. Ultimate responsibility for the proper consideration of wind type and direction in the structural design of the Citicorp Tower—as that design may have been revised—remained with the lead engineer. It is a responsibility that he recognized, accepted, and would ultimately bear.

The potential for harm was incalculable. To prevent that harm— and to heed the lessons of history—required only the close examination of a known danger.

10

"A SERIOUS AND DEADLY MATTER"

MONDAY, JULY 31

Following LeMessurier's induction to the National Academy of Engineering in February 1978, his stature in the industry skyrocketed, paving the way for opportunities around the world. He opened offices in Rome and accepted commissions in Singapore and Saudi Arabia. The merger of his engineering firm with Sippican in 1973, which he had so dreaded, had become wildly successful. "Things had blossomed," he recalled. "We were all getting rich."[1]

After completion of the Citicorp complex, LeMessurier was retained to design a performing arts center in Tehran and was shuttling between Iran and his office in Italy. One evening, following dinner, he suffered severe nausea and bouts of uncontrolled vomiting tainted with blood. He consulted with an American doctor in Rome who, after a thorough examination, concluded that there was no underlying issue and attributed his symptoms to severe indigestion.

Though weak, LeMessurier traveled the next day to Tehran, where his symptoms quickly worsened. Soon in total crisis, he was admitted into a private hospital with the help of his Iranian architect's wife. Once stabilized, he was transported back to Massachusetts, where he was diagnosed with Mallory-Weiss syndrome—a rupture of the esophagus. His doctors informed him that he had lost more than half of his blood and narrowly escaped death. "It made me realize my life had changed," LeMessurier recalled. "I was a different man as a result of this."[2]

Shortly thereafter, he was embroiled in the Citicorp Tower controversy.

◢

LeMessurier looked at Dorothy and then averted his eyes. There was much at stake: his business, his professional reputation, his financial future. He felt his world closing in, and yet, in his engineer's mind, there was a strange comfort in the certainty of the problem. He had assigned a statistical probability of the building's failure, and it left him with few realistic options. He explained to his wife the mathematical conclusions he had reached and gazed out over the waters of Sebago, as if to take in what he was about to lose. It was, he recalled, "a serious and deadly matter."[3]

He considered simply remaining silent. Alan Davenport and Al Romaneski knew what the problems were with the tower, but only he knew the full extent of those problems. Only he knew the *one-in-sixteen* probability of failure. Perhaps, he pondered, Davenport and Romaneski could be persuaded to loyally conceal what they knew. "I had information that nobody else in the world had," he recalled. "I had power in my hands to effect extraordinary events that only I could initiate."[4] But did he have the courage to exercise that power?

After LeMessurier's meeting of July 24 with his associates in New York, his oldest daughter, Claire, visited him at his hotel room. She was twenty-four at the time, living in the Lower East Side and working as a producer of a weekly cable-TV program on dance. Her father was distraught and had clearly been drinking. He confessed that he had made a terrible mistake with the Citicorp Tower and suggested—only half joking—that he should throw himself out the window. Given his emotional state, the comment worried Claire. Later at the lake house, confronted with the full gravity of his perceived error, LeMessurier once again considered suicide. "I did say to myself if I could drive down the Maine

Turnpike at a hundred miles an hour and deliberately drive into a bridge abutment, . . . that would be the end and all this would go away. I thought about that," he recalled.[5]

LeMessurier had never served in the armed forces, and he felt, perhaps, a twinge of guilt about it. Now, he reflected, was a chance to atone for past transgressions. He thought about his three young-adult children and how they might feel about him if the worst should happen with the building and they learned that their father had done nothing to prevent it.

In a moment of moral clarity and resolve, doubtless born of his recent life-altering illness, LeMessurier suddenly grasped his obligation to directly confront the Citicorp structural issue—and, perhaps, in the process, his own pangs of conscience. The problem, he almost gratefully recognized, was so sharply defined, so precisely delineated, that there was only one clear and unmistakable course to take. Regardless of the consequences, he must disclose the problem and do everything in his power to resolve it. "I didn't really care if all hell broke loose. If I did the right thing I'd at least have something to be proud of." He told Dorothy, "What happens to me professionally is something I can't allow myself to think about now, because I don't have a choice." It was, he recalled, "a peculiar mix of vanity, power, but almost joy, too."[6]

◢

Even as LeMessurier wrestled with structural flaws and ethical dilemmas, the Atlantic Ocean churned. Off the coast of Kill Devil Hills, just south of the beach where Orville Wright piloted the first heavier-than-air, motorized aircraft, a waterspout surged to life and barreled toward land. Howling fiercely, the seaborne tornado plowed into the Outer Banks resort community, tearing roofs from buildings, demolishing homes and cottages, and claiming the life of one elderly woman pinned beneath a hurled refrigerator. "I was so Godawful frightened out of my wits that I wasn't paying attention to the noise," said one witness.[7] Though waterspouts were fairly common in coastal areas, those with the destructive force of the July 31, 1978, event were generally rare—and, perhaps, a harbinger of the meteorological perils in the North Atlantic that lay ahead.

LeMessurier had considered, in a rudimentary way, how he might go about repairing the vulnerability with the Citicorp Tower; but as he drove south to Cambridge that Monday morning, almost certainly unaware of the events on North Carolina's Outer Banks, he knew what a hurricane would mean for the building, and he decided that he must immediately notify and mobilize his team. Time was in short supply.

When LeMessurier arrived at his office, he quickly brought Romaneski up-to-date with his analysis of the building's perceived vulnerabilities, and he placed a call to Hugh Stubbins, who he learned was on the West Coast and unavailable for discussion. Anxious not to delay the process, LeMessurier decided to contact Carl Sapers, Stubbins's lawyer. Sapers was a partner in the prestigious Boston law firm of Hill & Barlow and a preeminent expert in the fields of construction and architectural law. In 1975, he had been awarded the American Institute of Architects' Allied Professions Medal for excellence in the nondesign segment of the field.

Though LeMessurier knew Sapers well, the lawyer was hesitant at first to meet without his client Stubbins present, fearing a potential conflict of interest. LeMessurier persisted, giving him a brief overview of the problem and pressing him urgently on the need for swift action. A meeting, the engineer continued, was the best and only way to provide Sapers's client notice of the problem in his absence. At the same time, LeMessurier thought, Sapers could provide him with some much-needed guidance on how to proceed. Realizing the urgency of the moment, the lawyer agreed to meet LeMessurier at noon for lunch.

In what must have been a combination of humiliation and relief, LeMessurier presented Sapers with a full summary of the facts and an assessment of the potential risks to the Citicorp Tower in a quartering wind scenario. The lawyer listened intently and then asked whether LeMessurier's firm carried professional liability insurance to cover errors and omissions in the practice of its trade. With the stress of the discoveries LeMessurier had made in Ontario and Maine and the concern over how to handle them, he had nearly forgotten that he, indeed, had such a policy.

"Good," said Sapers blandly. "Call them and put them on notice."

Sapers advised LeMessurier not to speak with anyone else outside his own firm about the matter and not to take any unilateral steps to remediate the problem with the tower until the insurance company was notified, since to do so could void coverage. Then, he said, Citicorp was going to have to be informed.

Later that day, Romaneski, then the president of Sippican Consultants International, telephoned the claims handler for Northbrook Insurance Company and reported a "strengthening need" in the Citicorp Tower bracing system.[8] Confirming the conversation in a letter of the same day, Romaneski wrote, "there is a serious design error in the Citicorp building in New York City."[9]

11

REVELATIONS

"They thought I was nutty," recalled LeMessurier. "Nobody in his right mind calls up and says 'my building is going to fall [down].'"[1]

Perhaps for that reason, and others having to do with less "delicate" concerns, the lawyers representing Northbrook Insurance wanted an opportunity to personally evaluate LeMessurier and his apprehensions about the Citicorp Building. In a conference call on July 31, 1978, with Al Romaneski and various personnel from Northbrook, including its outside counsel, Max Edelman, it was agreed that LeMessurier would appear the following morning at Edelman's New York office to further explain the situation.

In 1978, Kroll, Edelman, Elser & Wilson was a law firm in transition. Sol Kroll, one of its original partners, had recently departed, and Max Edelman, the firm's founder, was creating a new and more expansive legal practice. While the firm was developing its client base and diversifying its realm of services, its core identity remained, at the time, centered on representing and defending insurance companies throughout the nation.

At eight o'clock on the night of July 31, an attorney from Edelman's office placed a call to the New York structural engineer Leslie Robertson, requesting his attendance at a meeting the following morning at the firm's office on Lexington Avenue. Given the nature of the practice and the gravity of the matters it often handled, an after-hours call from one of its lawyers was not unusual.

"What's it about?" asked Robertson.

"You'll find out when you get to the meeting," the lawyer said.

"Sorry," Robertson responded, "I have other things to do—I don't attend meetings on that basis."—and he hung up the phone.[2]

Moments later, another of the firm's lawyers, Roy E. Pomerantz, telephoned Robertson and more diplomatically explained that the proposed meeting involved the design of the Citicorp Building and would include William LeMessurier and lawyers for the Northbrook Insurance Company.[3] Robertson's presence to evaluate the issue would be of immense value, said Pomerantz. Sensing urgency and imperative, Robertson agreed to be there.

The lawyers had originally suggested another engineer whom LeMessurier had opposed due to the individual's limited experience with high-rise construction. Robertson, given his extensive background in New York City skyscrapers and in wind effects in particular, was a far more suitable consultant, LeMessurier told them. "I wanted to get the best critic of my own thoughts," he later said.[4] Upon quickly investigating Robertson's credentials, the lawyers agreed and sought his assistance.

Leslie E. Robertson would be described by his associates as "a genius" and "a perfect architect's engineer."[5] Over the course of his six-decade career, he provided consultation and structural design for a number of iconic high-rise buildings, including the US Steel Tower in Pittsburg, the Bank of China Tower in Hong Kong, the Puerta de Europa Towers in Madrid, and the Shanghai World Financial Center. It was, however, his first major commission, the Twin Towers of New York's World Trade Center, that would truly define his legacy.[6]

A native of Manhattan Beach in Southern California, Robertson always considered himself a "terrible student."[7] He dropped out of high school to join the Navy in World War II, where he served for a short time as an electronics engineer. Following the war, he was admitted by examination to the University of California, Berkeley, where he studied electrical and civil engineering. He was influenced by Berkeley's progressive ideals and adopted a philosophy of passivism and social responsibility that he embraced throughout his life.

Following college, Robertson held various positions at engineering offices on the West Coast before being selected by the Seattle-based structural design firm of Worthington-Skilling in 1963 to lead its New York City office. He would ultimately become a partner of the firm and remain in New York for much of his life.

Robertson would become an international lecturer and the recipient of four honorary doctorate degrees in engineering and science. Throughout his career, he amassed numerous accolades and distinctions, including recognition as the *Engineering News-Record* "Man of the Year" in 1989. His contributions to the World Trade Center project in New York earned him the Mayor's Award for Excellence in Science and Technology. Notably, he would become a leading authority in structural disaster prevention, fostering strong connections with politicians, engineers, physiologists, psychologists, and meteorologists nationwide—partners with whom he would often collaborate to analyze structural responses to earthquakes, floods, windstorms, and other natural disasters.

TUESDAY, AUGUST 1

Early on Tuesday morning, LeMessurier flew to New York and arrived at the offices of Kroll, Edelman, Elser & Wilson in the Graybar Building on Lexington Avenue, just eleven blocks south of the Citicorp Tower. Entering the "quintessential commercial colossus," as one historian described the Graybar Building, LeMessurier noticed the brick and limestone façade, rising in receding tiers garnished with stone reliefs that evoked the essence of Gotham-era New York.[8] It was conceived in 1925 to be the "largest office building in the world."[9]

Robertson had not yet arrived. The meeting had been scheduled for 8:45, but Pomerantz had asked him to wait a few hours so they could privately assess LeMessurier on their own. The situation was somewhat unusual for the lawyers. There had been no outside claim of injury or damage, and the Citicorp Building was visually sound and showed no perceptible signs of distress. LeMessurier's notification was an attempt to *prevent* a future harm that he had ostensibly created. In essence, he was blowing the whistle on himself.

"The whole problem is so simple if you think about the geometry," he began nervously. "It doesn't take more than 10 minutes to realize that this is a disaster."[10] He explained the difference between perpendicular winds and quartering winds and briefly described the effects of those conditions on conventional square and rectangular buildings such as the Citicorp Tower. He maintained that neither code nor practice mandated consideration of nonperpendicular winds in the design of high-rise structures and that such wind conditions were not widely discussed in the engineering literature or the classroom. He explained that he had, in fact, discovered the issue through a series of matter-of-fact calculations that he performed after a call from a random college student that piqued his curiosity.

The impact of quartering winds, he claimed, would not have been of any consequence had the original design of welded connections in the diagonal bracing system been adhered to. Once those connections were converted to bolts, however, a change he insisted he was unaware of at the time, it fell upon his engineers to accurately design the capacity and strength of those bolts. He then looked down and almost sheepishly admitted that his office had failed to consider nonperpendicular winds in that design. The result, he told them, was a series of vastly overstressed bracing connections in the tower.

LeMessurier had prepared a streamlined summary of his calculations, and he circulated it among the lawyers. The Citicorp Tower, he told them somberly, was in great danger.

Max Edelman peered quizzically at the engineer. He asked what strength wind would take the building down. Wind velocity, said LeMessurier, is not the proper gauge of structural vulnerability. Winds at the top of a structure may blow at different strengths and from different directions than at the bottom. Each structure, he continued, may have different sensitivities to different wind strengths.[11] Proper design-to specifications created by engineers, he explained, are based, instead, on statistics—the "hundred-year flood," the "five hundred-year flood," or the "ten-year earthquake." He looked sharply at Edelman and asserted that the Citicorp Tower had a fifty-fifty probability of failure in a storm that could hit New York City once every sixteen years. He then reminded the gathering that the hurricane season was fast approaching.

At 11 a.m. Leslie Robertson arrived, as did Stanley Goldstein of LeMessurier's New York office and Murray Shapiro and Leo Plofker from the Office of James Ruderman, each of whom had been requested to attend without knowledge of the exact nature of the meeting or the problem at hand.

LeMessurier once again provided his detailed narrative of the Citicorp Building design flaw and the vital need for quick corrective action. As LeMessurier anticipated—and to his immense relief—Robertson immediately grasped the severity of the issue and unequivocally confirmed to the gathering that, given the findings presented, they were, in all probability, facing the imminent prospect of total collapse of the tower in severe winds. The building, Robertson agreed, was in "immediate and grave danger."[12]

It was the confirmation of credibility that the firm's senior partners had sought. The lawyers now clearly understood that they were dealing with a crisis, the potential of which could reverberate through the insurance industry like a shock wave.

The meeting continued for much of the day, with Robertson, in the words of one commentator, "tall, trim, fluent, humorous, quick-witted, blithe, . . . but there's that enormous ego," and LeMessurier, "older by two years, . . . voluble and intense with a courtly rhetorical style," each taking center stage.[13]

LeMessurier knew instinctively that he could not disclose the structural issue without having at least a rudimentary plan to fix it. While in Maine, he had developed such a plan. He told the gathering that during the design phase of the tower, Hugh Stubbins had resisted the idea of placing the diagonal bracing system on the exterior of the building. Instead, the braces had been placed just inside the exterior glass and behind the interior sheetrock walls. The connections, therefore, were easily accessible for repair. All that was needed, he explained, was a series of six-foot-long steel plates—or brackets—welded over the existing joints to reinforce the deficient bolted system.

The engineers had taken comfort in the fact that any storm with enough destructive force to significantly impact the building should come with adequate warning. Weather could be monitored, and hurricanes could be tracked. None of them felt that an evacuation of the building was warranted at the time, but Robertson pointed out that in his experience, the bank, once advised of the problem, may be likely to overreact and order that the tower be cleared of occupants.

"I almost fainted," recalled LeMessurier.[14] He was convinced that evacuation was the step most likely to assure the end of his career. He explained to the group that the TMD would certainly provide some level of protection against a catastrophe and that emergency generators could be installed to ensure its continued operation.

Robertson was having none it. He argued to the lawyers that the prudent and most conservative course was to assume that the damper would become disabled in a massive storm and that the building should therefore live or die on its own structural merit. For this reason, urged Robertson, any curative and safety measures must be undertaken without reliance on the TMD.

It was the first hint of friction between the two engineers. LeMessurier believed that Robertson was being overly cautious and that the insurance company and Citicorp would, as a result, be coaxed into measures that may not be required. "Robertson, at that very meeting, and this became characteristic of him from then on, was a real alarmist," LeMessurier later said, "a professional worrywart."[15]

The two men, no doubt, shared a mutual respect for each other. Robertson admired LeMessurier for being forthright with the Citicorp Building issues, and he complimented him for his composed and in-charge demeanor throughout the crisis. And LeMessurier esteemed Robertson as a giant in the field of structural engineering and for his knowledge of wind effects on high-rise structures. But their personalities and working styles generally clashed. Robertson, by nature, immersed himself in the details of engineering, while LeMessurier tended to be more broadly and conceptually involved in his projects. "Bill is not . . . a detail kind of thinker," Robertson later noted. "There are detail thinkers in the firm, . . . but Bill himself is temperamentally not that kind of person. . . . That's one of the huge differences between Bill and me, I'm a nitty gritty kind of guy and he's not."[16]

"It's a bittersweet relationship we had," said LeMessurier.[17] Robertson agreed. "It would be improper to say that Bill and I saw everything eye to eye. . . . [We] are both pretty strong-willed individuals. . . . We don't pull punches, don't play politics."[18]

Despite the two men's obvious differences in style and opinion, as the meeting drew to a close, they agreed, and the lawyers directed, that LeMessurier would immediately inform Hugh Stubbins of the problem with the Citicorp Tower and that the engineer and architect would, together, notify the owner as soon as possible.

◢

The Citicorp Center was Hugh Stubbins's magnum opus—the masterpiece of his professional career. He was tremendously proud of the building and the design he had conceived. Though he and LeMessurier had been friends and business associates for years, it was with trepidation that LeMessurier, in the company of Carl Sapers, knocked at the door of the architect's home in Cambridge late that evening, following his return from California.

Stubbins "was a bit shocked" after being advised of the news, he would later admit. "I'm not sure that the seriousness of it penetrated all the way at that moment."[19] Stubbins's associate Easley Hamner was less circumspect when his boss ultimately filled him in. "It was probably the scariest moment of my entire life," he recalled.[20]

The men talked for several hours and agreed on a strategy for repairing the problem. That was the easy part. The more vital and delicate task, they recognized, was to inform the bank's senior officers, as gently as possible, that their new office tower was in danger of collapse.

LeMessurier had established few contacts with Citicorp's upper management during the design and construction of the tower. His client was the architect, not the owner. Stubbins, however, often dealt directly with the bank's chairman, Walter Wriston. If action was to be taken, said LeMessurier, they needed to go straight to the top.

LeMessurier told Stubbins, "We're going to need your help, Hugh. . . . I [already] got a plane ticket for you: first class."[21]

WEDNESDAY, AUGUST 2

As LeMessurier and Stubbins shuttled back to New York, Leslie Robertson was already meeting with Stanley Goldstein at LeMessurier's Madison Avenue office, reviewing the Citicorp drawings and testing the firm's wind-load calculations. In scrutinizing the plans, Robertson immediately recognized an additional problem with the lateral bracing of the tower's chevron floors. Above and beyond the issue of the bolted connections and their vulnerability to quartering winds, the building, in Robertson's view, lacked an adequate floor diaphragm design or necessary support structures for the floor-beam connections.

It was information that LeMessurier was not in the mood to receive. As he and Stubbins arrived at the office, he explained to Robertson that his first priority was to notify the bank of the condition of its building. At his urging, their focus accordingly shifted to the development of a preliminary plan of essential actions to be communicated to Citicorp management.

"You don't call Walter Wriston and expect him to jump for a meeting in the next twenty minutes," LeMessurier later joked. "He was the most powerful financial person in the world."[22] Nonetheless, he and Stubbins tried frantically to reach Wriston on the phone. After several failed attempts, they tried to contact the bank's president, William Spencer, also to no avail.

LeMessurier then recalled that John Reed, a Citicorp senior vice president and graduate of MIT's Sloan School of Management, had been very helpful with the implementation of the tower's TMD. He had worked with Reed on that phase of the project and found him to be cool-headed, approachable, and conversant with basic engineering concepts.[23] And more importantly, Reed turned out to be much more accessible than his superiors.

At 10:30 a.m., LeMessurier and Stubbins were seated in Reed's office. LeMessurier told the executive that through "an unusual chain of circumstances," he had discovered a flaw in the Citicorp Tower.[24] He described the dangers presented by quartering winds and explained the statistical probability of a critical-strength storm approaching New York within the next sixteen years. Though the TMD, retrofitted for auxiliary power, would definitely provide added protection, its operability during extreme conditions could not be relied on with absolute certainty, he warned.

Remembered by one writer as "a cocky, impish technocrat, . . . once considered the boy wonder of American banking," Reed silently appraised the two men.[25] "[He] remained calm, sober, absolutely expressionless," recalled LeMessurier.[26] The problem, LeMessurier continued, could be corrected efficiently and inexpensively by welding steel splice plates to the bracing joints, but with the September hurricane season fast approaching, urgent action was required. He exhorted Reed to begin mobilization of a management team at the bank at once to implement the corrections.

LeMessurier emphasized that he and Stubbins were in New York for the sole purpose of addressing this problem, and he reminded Reed that he had a satellite office just down the street. Reed, assuming that he would have to report to the chairman before taking any substantive action of such magnitude, instructed them to carry on with their day and await further word.

After an hour of impatient unease and detailed discussions with Leslie Robertson, the two men decided to have lunch at a restaurant across the street from LeMessurier's office. LeMessurier told his secretary of their plans and instructed her to interrupt them if anyone from Citicorp called.

Robertson stayed behind and focused his attention on procuring advanced warning systems for structural shifting and weather. Drawing from his broad array of contacts and resources in the field of emergency preparedness, he called a California company that he had dealt with in the past, called Kinemetrics, for the placement of strain gauges to provide monitoring of the stresses on the tower's chevron bracing system until the structural repairs could be completed. Robertson supplied his contact, Lee Benuska, the company's vice president, with detailed specifications of the tower, and he outlined the design flaw they were facing. Benuska promised that, if engaged by the bank, he could quickly deliver seventy-five strain gauges directly to the building and have a technician on-site right away.

Next, Robertson made inquiries with weather forecasters and a meteorologist with whom he was acquainted for the provision of weather data and advance storm warnings.

At about 1:30 p.m., as LeMessurier and Stubbins finished their lunch, LeMessurier's secretary appeared suddenly at their table. Walter Wriston would meet them at their office in ten minutes, she told them breathlessly.

◢

In an era of "genteel, stodgy, and risk averse bankers," Walter Wriston brought a new and visionary entrepreneurial spirit to the financial services industry.[27] Born in Connecticut and raised in a strict midwestern Methodist home, Wriston graduated from Wesleyan and earned a master's degree from the Fletcher School of International Law and Diplomacy. Following military service in the Signal Corp in the Philippines during World War II, he joined First National City Bank as a junior comptroller. By 1967, he was president and CEO of the company and, two years later, chairman of its successor, Citicorp, a position he held until 1984.

"In those seventeen years," wrote Phillip L. Zweig, Wriston's biographer, "Wriston . . . built Citicorp into the world's mightiest banking power and established for himself an unchallenged reputation as the world's most influential banker."[28] Under his tenure, Citicorp expanded across state lines and into global markets, thrashing regulatory barriers and bringing technology into the forefront of the industry. He introduced financial innovations such as certificates of deposit, automated teller machines, and the Master Charge credit card system. Through Wriston's leadership, Citicorp enjoyed an unprecedented rise in annual profits from $145 million at the beginning of his tenure to $890 million by his retirement.[29] "By betting on youth and brains," wrote Zweig, "he revolutionized the way Americans managed their money—and built a money machine for Citicorp in the process."[30]

Wriston's influence would extend beyond banking and into the political sphere. Staunchly conservative and politically allied, he served as an economic adviser to the Nixon, Ford, and Reagan administrations, where his insights and expertise were instrumental in shaping economic policy. He would become chairman of President Reagan's Economic Policy Advisory Board, where his counsel contributed to the deregulation and free-market principles that became hallmarks of the economic policies of that era. On two occasions, Wriston said, he turned down offers for the position of Treasury secretary of the United States.[31] In June 2004, President George W. Bush awarded Wriston the Presidential Medal of Freedom, the nation's highest civilian honor.

A towering presence, Wriston entered the office accompanied by John Reed and immediately seized the attention and respect of the gathered group. "I thought he was ten feet tall," remembered Robertson. "I was so impressed with that man."[32]

Stubbins, who had met Wriston several times during the design phase of the building, introduced everyone. LeMessurier, still wary that he might be perceived as somewhat of a "nut," quickly pointed out that Robertson, the structural engineer for the World Trade Center, had agreed to provide independent guidance on the Citicorp Tower. Even Robertson later admitted to a feeling of being evaluated.[33]

LeMessurier apprehensively began recounting the problem with the Citicorp Building and its precarious structural state. He described the issue with the bolted connections and the failure of his office to adequately consider nonperpendicular winds in the design specifications for those bolts. The room was filled with design professionals, and the issues discussed were technical in nature. Wriston remained stoic and a bit puzzled. Sensing that LeMessurier may not have conveyed the full scope and gravity of the problem in an understandable manner, Robertson looked squarely at the chairman, grabbed a clipboard from the table, and placed it on its end. "Gentlemen," he began, "if the wind blows"—he slammed the clipboard down on its face with a loud thud—"that's what will happen to your building."[34]

From that moment, the tone of the discussion shifted from explanation to action. Robertson and LeMessurier had agreed on a preliminary strategy for the strengthening of the wind-bracing connections by incorporating the remedial steel plates and, in the interim, to enhance the operational reliability of the TMD, to develop a weather-prediction warning system, and to equip the building with strain-gauge monitors. Those were the essential points of action, said the engineers. Robertson added that preliminary discussions with several providers of services were already under way.

To the men's relief, Wriston immediately authorized implementation of the strategy, and he committed the full support and resources of the bank to the effort. He recognized that his personal role in the process would be less about technical details and more about public relations. As the meeting progressed, he began considering the initial phrasing of a press release and how on earth he would explain to the tenants of the building the flurry of activity that was about to come. He asked for a yellow pad to write down his thoughts and said through a burst of laughter, "All wars are won by generals writing on yellow pads."[35]

"It took a big load off of my mind to have that kind of reaction," recalled Stubbins. "They're not blaming anybody."[36]

◢

Within hours of the meeting, Wriston had mobilized his forces. That afternoon, LeMessurier hosted a further conference attended by the bank's upper management team, including John Reed; William Spencer; Citicorp's in-house counsel, Hans Angermueller; and vice president Robert Dexter, whom Wriston had appointed as the bank's contracting officer for the remedial work. Stubbins, Robertson, Stanley Goldstein, and soon Al Romaneski, who had flown in from Boston earlier in the day, also joined the gathering. As they coordinated the strategic details of shoring and monitoring the building, they reached consensus for chain of command and the establishment of communication channels. Hal DeFord and Robert Dexter, who had worked extensively with the design team during construction of the building, were once again pressed into action as primary liaisons for the repair program. The engineers, it was agreed, would have general decision-making autonomy, but ultimate contracting decisions would be left to their client. To address the need for rapid contact and response, DeFord and Dexter arranged for emergency pagers to be provided to each member of the bank's upper management as well as Stubbins and the structural engineers. It was agreed that all hands would be on deck.

Romaneski went right to work. He called Bill Kasel from MTS Systems Corp. in Minneapolis, the manufacturer of the TMD, and urged him to come to New York right away. Emergency structural improvements to the tower were in the works, Romaneski said, and his team required consultation on the capacity and operation of the damping system.

At the same time, LeMessurier contacted Joseph Loring, who had provided mechanical services in the construction of the Citicorp Center and whom LeMessurier regarded as "one of the great electrical engineers in New York."[37] LeMessurier briefed Loring on the structural problems with the building and asked him to begin work on an emergency generating system to keep the TMD powered during extreme weather.

All through the day, LeMessurier kept Max Edelman informed of the actions being taken and was pleased with the progress he could report. The meetings, he said, had succeeded in establishing transparent, candid, and amicable relations with Citicorp's senior management and what LeMessurier called a "spirit of mutual trust."[38]

He could only hope it would endure.

12

MOBILIZATION

After a monthlong appearance at the Wedgewood Room of New York's Waldorf Astoria Hotel in the summer of 1944, Frank Sinatra became a regular patron and performer at the hotel. He would continue his association with the Waldorf throughout much of his professional life and, while married to Mia Farrow in the mid-1960s, took up residence in suite 2500 on the twenty-fifth floor. Sinatra had always loved the posh, forty-three-hundred-square-foot, five-bedroom penthouse on the thirty-third floor, owned by Cole Porter, and when it finally became available in 1979, he and his fourth wife, Barbara, moved into the apartment as residents of The Towers at the Waldorf for the princely sum of $1 million per year. Porter had called the ten-room suite a "dream of beauty."[1]

As the Citicorp crisis deepened, LeMessurier decided that he needed centrally located head-quarters—a "war room" of sorts—to host consultants and conduct meetings in connection with the tower's structural renovations. "I realized I needed a whole office night and day—and a place to live."[2] He had been staying at the St. Regis, but the room was not large enough for his purposes. He learned that the cousin of one of his New York employees was the night manager of the Waldorf, and by Thursday morning, he had arranged a long-term rental of Sinatra's twenty-fifth-floor suite while the singer was on tour in Los Angeles and other locales. "Everything [was] painted green and coral," he recalled.[3] With its multiple bedrooms and expansive living space, the suite was perfect for LeMessurier's business and personal needs, décor notwithstanding. Stubbins would later call the space their "Disaster Headquarters."[4]

That summer, "Old Blue Eyes" began each concert with his rendition of "Night and Day"—perhaps a tip of the hat to the earlier resident of suite 33A, which he was already in negotiations to rent.

THURSDAY, AUGUST 3

"Things come thick and fast over the next week," LeMessurier recalled.[5] On that Thursday morning, Les Robertson telephoned Karl Koch, a New Jersey steel fabricator, and told him he needed evaluation of some remedial work at the Citicorp Tower in New York City. Karl Koch Erecting Company had provided the steel for the World Trade Center, and the two men knew each other well; but Koch balked. They were backed up with work, he said, and it would be difficult to take on a project of this size. Robertson persisted, claiming that the situation was critical. He then detailed the structural problem that LeMessurier had discovered, and he urged Koch to assist. The tower's wind-bracing system needed to be retrofitted with reinforced steel plating, Robertson explained, and the work had to be done as quickly as possible to avert a potential catastrophe. Koch admitted that he did have a ready supply of two-inch-thick, high-strength steel slabs in stock that he could possibly make available to Robertson but cautioned that fully licensed welders with high-rise experience would be required for its installation.

Bill Kasel arrived in New York and immediately received a full briefing by the engineers. He warned the team that the TMD had been designed to *deactivate* under certain extreme conditions. The device, he explained, was never intended to rescue the building from disaster, only to stabilize it for the comfort of its occupants during excessive building sway. He added, however, that his company might be able to make some temporary adjustments to the machine's controls to sustain its operation until the crisis was over.

Later that morning, Les Robertson, Bill Kasel, the LeMessurier-Ruderman engineering team, and two Koch engineers met with DeFord and Dexter at the Citicorp Tower. In an empty office on the thirtieth floor, several of the bank's custodians tore away a section of drywall, revealing a diagonal splice area typical of that found throughout the structure. The group peered at the massive steel members and the four bolts that connected them, each man pondering the required repairs and his personal role in the undertaking.

The Koch engineers took measurements and snapped several photographs. They reiterated that certified, licensed, preferably union welders would be critical to complete the corrective work. LeMessurier studied the exposed connection, privately embarrassed by his role in its faulty design. He likewise measured the splice area and discussed with the engineers the anticipated specifications and positioning of the required steel plates. He promised that his team would immediately begin design of these reinforcing brackets, to be fabricated by Koch.

DeFord and Dexter then accompanied Kasel and the structural engineers to the upper floors of the building to view the TMD. LeMessurier had pinned his personal hopes on the device, and his pulse quickened as they entered the housing room. Kasel inspected the system and told the group that it seemed to be operating properly. He promised to quickly contact Joe Loring to assist with the installation of a backup power generator, and he pledged twenty-four-hour-a-day on-site MTS technical staff to monitor and service the machine during renovations. LeMessurier sighed with relief. "Those guys charged like bandits from Minneapolis," he recalled. "This wasn't their fault. Suddenly you have . . . experts on that machine, with spare parts and lubricants and chips and anything else that might be needed. . . . That was my big security blanket, that's how I got through the crisis."[6]

LeMessurier was not going to make the same mistake twice. He spoke at length with Robertson about the design specifications for the reinforced steel plating, adamant that it would be correctly engineered.

Throughout the day and late into the night, LeMessurier and his associates huddled in the Waldorf suite recalculating expected wind loads and developing the most suitable design requirements for the reinforced plating. "In essence," he later wrote, "a very conservative policy was established to cover any new information yet to be developed by Davenport at the wind tunnel in Ontario."[7] On the basis of the specific criteria worked out by the team, Stanley Goldstein was tasked with the formulation of final, detailed drawings to be utilized by Koch in the fabrication of the curative steel plates.

That morning, the bank mobilized a local contractor to quietly begin unveiling certain diagonal splice areas of the building and to begin construction of plywood welding booths in preparation for the corrective repairs. Citicorp had not yet transitioned its offices from 399 Park Avenue, and a full third of the building still remained unoccupied, allowing for a degree of secrecy.

As Robertson had recommended, DeFord and Dexter approved the engagement of Kinemetrics for the strain-monitoring instrumentation and authorized him to employ weather-service personnel to establish early warnings of destructive winds. Robertson had first attempted to retain the National Weather Service as his primary source of wind and weather information but was unable to establish fruitful communications. He then turned to the National Weather Corporation of Newark, New Jersey, a private forecasting company, to provide required weather data. At an 11 a.m. meeting at Robertson's office on Park Avenue, the president of the company, John Woolley, promised a collection of fourteen forecasters at Newark International Airport and two additional forecasters at Westchester County Airport to provide a minimum of twenty-four-hour advance notification of any winds in excess of seventy miles per hour that were part of a major weather front and four to six hours' notice for locally developed storms.[8]

"Not good enough," said Robertson. Uncertain of the exact level of the tower's vulnerability, he insisted on predictions of winds in the New York metropolitan area over forty-five miles per hour of more than five minutes in duration, including wind direction. After some further equivocation, Woolley hesitatingly agreed to Robertson's terms, confident that he could fulfill the promise but knowing that he would need to draw on more resources than usual to do so. Woolley also agreed that his office would transmit regular reports by telecopier to both Robertson and Citicorp management.

Irving A. Singer, a local meteorologist who was "long familiar" to Robertson, also attended the meeting. Weather prognostication was not Robertson's forte, and he wanted someone available to explain in plain English the information generated by National Weather. The men agreed that Singer would act as a liaison between the forecaster and the engineers, assisting in the "analysis, interpretation, and quality control of the data."[9]

An hour later, Robertson convened another meeting at his office with DeFord, Dexter, LeMessurier, and Romaneski. They discussed the progress of the operation, and Robertson informed the group of his weather-alert arrangements. Then, the conversation shifted to the efficacy of the TMD. The contractor, MTS Systems, had begun full-time surveillance of the device, and work was already under way to provide emergency power generation. Romaneski pointed out that the company's technicians were in the process of evaluating measures to increase its performance, and any day, the California company, Kinemetrics, would be installing instrumentation to fully monitor the stresses on the building.

Robertson remained unmoved regarding the TMD. He explained to the gathering his continued concern that electrical backup systems could also fail and should not be relied on as a fail-safe during repairs. Yes, the building could be monitored, and forecasters could predict most weather events; but it was the freak storm, the rogue gust, the unforeseeable occurrence that no warning system or stress gauge could ever prevent that most troubled the engineer. The structure, he told them, should be evaluated in the most conservative light, on the basis of its existing engineered state, without dependence on damping protection.[10] As a result, he said, plans should be developed in conjunction with Washington, DC, consultants with whom he was familiar for the evacuation of the building and the surrounding areas when and if conditions warranted. LeMessurier traded glances with DeFord and Dexter. Development of an evacuation plan for the displacement of thousands of people was far beyond the bank's managerial experience, and now Robertson was suggesting that they contact national disaster-control experts.

Until then, only a small and restricted inner network of people, bound by confidentiality, had knowledge of the Citicorp Tower's precarious situation. DeFord and Dexter had been hesitant to expand that network, fearing the consequences of broad disclosure for the company and the city. That network, however, was about to dramatically expand.

MOBILIZATION

13

"ANGEL OF THE BATTLEFIELD"

MONDAY, AUGUST 7

LeMessurier had traveled to Maine Friday night to see his wife and gather his thoughts. He was relieved that his disclosures about the building had been largely met with cooperation and goodwill and that planning of the remedial work was progressing smoothly. Sweeping challenges remained, however. He privately scorned Robertson's urging of evacuation plans but wisely decided to remain reticent on the issue and to distance himself from those decisions. They were not his to make. He returned to the New York "War Room" on Sunday to join Stanley Goldstein, who worked through the weekend generating the final drawings for the corrective steel plates.

Confident in the precision of the new splice design plans, LeMessurier submitted them for examination in the early morning of Monday, August 7, to Arthur Nusbaum, project manager of HRH Construction, the original Citicorp general contractor that was once again commissioned to spearhead the remedial welding work. After review and some questions, Nusbaum, in turn, furnished the plans to Karl Koch Erecting for final fabrication.

LeMessurier then made his way to Citicorp headquarters for further discussion on the status of repair strategies and the unsettling requirement of evacuation planning. In the office of the bank's general counsel, Hans Angermueller, Robertson suggested to the gathering of Citicorp officers as the first order of business that he notify city building officials of the structural problem and the contemplated remedial strategy. LeMessurier readily agreed and realized that he had been so preoccupied with other details that he had completely forgotten about involving the local government. Robert White, a prior Ford Motor Company executive hired by John Reed to streamline check-clearing systems for the bank, now replaced his boss as executive vice president in overall charge of the Citicorp building crisis and consented, together with DeFord and Dexter, to informing the city. Robertson promised to quickly initiate the necessary communications.

The conversation then shifted to evacuation planning.

The group first evaluated the tenure and qualifications of Citicorp's existing security staff, including its protection officer, deputy protection officer, and protection specialist, all of whom had some degree of training and experience in emergency preparedness. The threat of domestic and international kidnapping and extortion plots against financial institutions such as Citicorp had become a known industry hazard, and long-standing contingency plans had been developed to deal with such incidents. But private security came with obvious shortcomings. Knowing that the bank would have to look beyond its internal resources to deal with the unthinkable possibility of structural collapse, Robertson had already initiated preliminary contacts with national evacuation experts with whom he had collaborated in the past.

As much as DeFord and Dexter dreaded the corporate black eye that would surely come with an evacuation of the Citicorp Building and its surrounding neighborhoods, they recognized that doing nothing in the face of advanced knowledge would be worse—if not criminal.

And recent history weighed heavily on their minds.[1]

◢

At 8:37 on the evening of July 13, 1977, a lightning strike near Westchester County, New York, caused the Indian Point Energy

Center to disconnect from the electrical grid, triggering a cascading sequence of events that eventually led to the failure and total collapse of the Consolidated Edison power system, plunging the entire city of New York and surrounding areas into darkness. According to an impact assessment of the occurrence, "A combination of natural phenomena, improperly operating protective devices, inadequate presentation of data to the system dispatcher, and communication difficulties" all contributed to the calamitous event."[2] In what would be remembered as one of the nation's most devastating power failures, nine million people huddled in darkness for over twenty-four hours.[3]

During the outage, the city exploded in a wave of plunder, arson, and destruction, worsened by the region's existing economic woes. A combination of public and private emergency services worked tirelessly to provide relief and support to the affected communities, yet many still suffered. The blackout underscored the frailties of New York's disaster readiness and would become a catalyst for improvement of the city's emergency response protocols.

DeFord and Dexter vividly recalled the 1977 power outage and its lessons for emergency preparedness. If mere darkness could throw the city into turmoil, what would a toppling skyscraper evoke? they wondered. Other, less foreboding disruptions earlier in New York's history also proved instructive to the men.

By the expiration of Transport Workers Union Local 100's contract with the New York City Transit Authority on January 1, 1966, the union, led by its fiery boss, Mike Quill, had clearly had enough. On the first day of New York City Mayor John Lindsay's term of office, Quill and thirty thousand of his union members illegally walked off the job, leaving millions of the city's commuters without bus or subway service. "It is an act of defiance against 8 million people," Lindsay said angrily.[4]

With much of New York City's population fully dependent on public transportation, the strike would paralyze the city. Schools were empty, businesses were shuttered, and hospitals and other institutions operated with skeletal staffing. On January 2, a New

York judge granted an injunction against the striking workers and the following day issued an order for Quill's arrest. "The judge can drop dead in his black robes," Quill said upon hearing the news.[5]

After twelve rancorous days, the strike was finally resolved with a mediated settlement granting transit workers increased pay, pension improvements, and other paid benefits. Quill, who was in poor health at the time, would not survive the month, but his legacy—and that of the 1966 transit strike that he spearheaded—lived on as a watershed moment in the history of New York City's labor movement.

The 1966 strike was remembered also, however, as a test of emergency readiness. Contingency planning instituted by private companies and public institutions helped to reduce the financial blow caused by the work stoppage and provided lessons for the alleviation of future labor action impacts. In the Citicorp officers' review of their own records, they noted that during the 1966 transit strike, implementation of their internal planning activities resulted in "facilitating effective movement of . . . staff."[6] To Citicorp, the critical lesson of the 1966 transit strike was to quickly assume responsibility for its geographic area and to understand the needs of the affected population.

The ongoing threat of a New York City Police Department work stoppage, such as the five-day "blue flu" that took place in January 1971, presented an additional security concern for the bank and provided its management team with further context for its current emergency preparations. As a major financial institution with a substantial presence in the city, the company was required to consider alternative provisions for the protection of staff and patrons in the event of a police strike. In 1971, the bank formed a planning committee with the goal of enhancing private security resources in instances in which the police department could not be relied on. Its buildings had to be secured and it patrons safeguarded.

The newly constructed Citicorp Center, with its quasi-public outdoor facilities and its massive buildings populated daily by thousands of tenants, occupants, and patrons, presented an even greater security challenge. The consequences of inadequate arrangements in the absence of suitable police protection were obvious and potentially perilous. The bank's officers connected the 1971 public safety concerns to the current crisis and recognized their responsibility to the building's occupants and to the community as a whole. The threats were different, but the obligation was starkly similar.

Evaluation of international crises also provided Citicorp with context for the implementation of its emergency response. On February 4, 1976, an earthquake registering 7.5 on the Richter scale rocked the Central American country of Guatemala. The event, one of the worst in the region's history, resulted in a reported twenty-two thousand deaths, seventy-five thousand injuries, and more than one million displaced residents.[7] Of the earthquake's destructive impact, one United States senator said in a subcommittee hearing on relief efforts, "It is staggering, literally staggering. We have always associated such a catastrophe as something that only the ultimate nuclear conflict could produce. Here it is in real terms, in very earthly context."[8]

Within Guatemala City itself, surprisingly few major problems involving emergency response were reported in the aftermath of the disaster, with the exception of the hospital system, which was dramatically impacted. "Fire services, police, military forces, telephone, water, power, and other similar services suffered relatively little damage and were able to respond effectively, at least in and around the city," wrote one researcher.[9] Overtaxed hospitals within the city, however, suffered damage to equipment and experienced the rapid exhaustion of supplies, making care of the overwhelming number of injured difficult at best.

Beyond the city limits, "transportation, communications, utilities, and emergency services in general were massively disrupted, . . . [but], confounding a popular myth, . . . public order did not dissolve and there was little panic or looting."[10] The most difficult problem faced in these areas was isolation. Damage to already crumbling roads and infrastructure made the provision of aid outside Guatemala City difficult and, at times, virtually impossible. Medical supplies, food, water, and other aid had to be airlifted where helicopters and small aircraft were available.

The international relief effort to Guatemala, coordinated through the United Nations, individual countries, and private charitable organizations, was, in the words of several observers, "truly staggering."[11] In some cases, the flood of goods, however well intentioned, resulted in more of a problem than a solution. Cultural indifference to the needs and tastes of the local population, together with the exorbitant costs of collecting, sorting, and distributing these goods, resulted in chaos and waste. Internationally provided medical supplies were frequently delivered in unusable condition.

The most critical need of the Guatemalan people in the aftermath of the 1976 earthquake, however, was shelter. With one million homeless citizens, the prospect of illness, injury, and further death became an appalling reality. The Guatemalan government maintained few housing initiatives prior to the earthquake, but evacuation and shelter of its people would quickly become its most urgent priority, particularly as the rainy seasons loomed.

In evaluating these real-life accounts and the level of advanced planning and preparedness demonstrated—and sometimes neglected—DeFord and Dexter plainly understood that the ramifications of a potential building collapse in New York City went far beyond public relations. Lives and livelihoods were at stake, and preparation for worst-case scenarios had to be undertaken.

While local fire and building codes provided some safety protocols, in 1978 federal guidance and support on evacuation planning was, at best, disjointed and regionalized. The Federal Emergency Management Agency (FEMA) was still in the planning stage, and disaster preparedness was often left to local governments and private business owners. When it came to developing evacuation plans, securing emergency funding, and assembling safety personnel, Citicorp could work with state and local officials but was essentially on its own.

After reviewing the crisis management case histories and further debating the merits of Robertson's proposal for federal evacuation experts, the Citicorp officers made the decision to consult the obvious and, perhaps, overlooked choice: the American Red Cross.

◢

While Angermueller, DeFord, and Dexter haggled with the structural engineers about evacuation planning, the Citicorp Tower was quietly placed on life support.

On the morning of August 7, a Kinemetrics technician began installation of strain gauges throughout the building for the purpose of detecting actual forces on the most susceptible elements of the building and to provide real-time alerts of any dangerous stress levels on those elements. These devices, together with amplifying circuits and balancing bridges to nullify interference of unrelated external forces, were welded to several of the main diagonals of the bracing system. According to Robertson, "[The] data could then be correlated with predictions based on wind speeds and . . . would allow at least a few moments warning in case of a dangerous level of response of the building to wind-induced excitation."[12] The actual warning apparatus—what Robertson called the "critical path" of the system—was a cluster of circuitry hardwired into the building and connected through existing telephone lines to a series of amplifiers and recorders mounted in Robertson's office at 230 Park Avenue and manned day and night. It was, according to LeMessurier, "a perfect way to measure the forces in the building at any instance."[13]

Kinemetrics, using its own nonunion electricians, was able to assemble and install the instrument package in record time. With few exceptions, the system would provide round-the-clock monitoring of the tower and real-time data on its soundness and performance. "Every time that building twitched we knew about it," Robertson later remarked.[14]

Meanwhile, on the nine upper floors of the building, the MTS technicians who had been closely monitoring the TMD now began beefing up the device. To improve its reliability during periods of severe weather, they increased the oil supply to the nitrogen actuators, installed larger supply-line filters to reduce oil contamination, and added a temporary hand valve to the bearing dump solenoid as a means to reduce built-up pressure. In order to avoid an unwarranted deactivation of the system and to increase the likelihood of its continued operation when needed, MTS also temporarily altered the original system settings of the machine. The technicians reduced the overall damping efficiency to 20 percent, raised the acceptable movement range of the structure—the so-called electrical stroke—to plus or minus fifty-one inches, doubled the allowable error-detection range, and bypassed a variety of noncritical temperature switches and fail-safes.[15] In essence, MTS was buttressing the machine's calibration as one might alter the "tilt" sensitivity setting on a pinball machine to tolerate conditions that would ordinarily cause deactivation in order to ensure its continued operation during extreme weather conditions. They were sacrificing durability for performance.

The management of MTS knew they were in uncharted territory. In a letter to the bank enumerating the required work and confirming authorization to proceed, the vice president of the company, R. W. Clarke, wrote, "Since we are modifying the TMD from a comfort device to a device that affects the structural integrity of the building, legal counsel has advised us that it is mandatory that you incorporate a full hold harmless clause in your written response . . . from and against . . . injuries to persons, including accidental death or property damage on account of . . . reduced comfort or occupancy of Citicorp Center."[16]

As Robertson had also reasoned, it was the unpredictable risks that unnerved MTS's management team.[17]

⊿

The National Weather Corporation, retained by Robertson to provide meteorological data, was already nervously plotting a circulating depression about 250 miles southeast of Brownsville,

Texas. Formed several days earlier from a cold front over northeastern Georgia, the storm brewed throughout the weekend, and by Sunday morning, an Air Force reconnaissance plane confirmed tropical storm strength. Until Bess, as the storm was named, unexpectedly began tracking southward, Robertson's forecasters believed that it might intensify into a marginal-strength hurricane. As it made landfall near Nautla, Mexico, on the morning of August 8, however, its rainfall quickly decreased, and the storm eventually dissipated.[18]

⊿

Nearly everyone involved with the Citicorp Tower repair project was working overtime. Endless decisions ranging from work authorizations to emergency preparedness had to be made, and every day the engineers and bank officers convened, conferred, and deliberated, often well into the night. In the background lurked a gaggle of insurance people and lawyers who had mercifully agreed to remain uninvolved in the details of the remedial project while the engineers and contractors completed their work. CNA, Stubbins's insurer, employed an independent structural engineer to monitor the crisis and advise of progress, and even Northbrook, LeMessurier's own liability carrier, retained another engineering consultant for unbiased guidance on the situation. In between meetings with bankers, consultants, contractors, and engineers, LeMessurier remained in close contact with Max Edelman to advise him of operational details and, ultimately, to discuss projected costs and potential liabilities.

From the beginning of Robertson's involvement in the project, there was some question about whom he actually served. During the first several days, it was agreed that he would act as LeMessurier's consultant. It quickly became evident, however, that the bank would also need its own independent technical guidance in the matter. Robertson raised the issue with LeMessurier and suggested that rather than bringing yet another engineer into the loop, it would be smoother if LeMessurier simply released him to the service of Citicorp. At first, LeMessurier hesitated. Was Robertson cynically attempting to align himself with the deep-pocketed bank rather than the engineer who potentially bore responsibility for the structural mistake? LeMessurier was disconcerted by the possibility, yet, despite their differences, he trusted Robertson and believed him to be ethically and professionally responsible. He shuddered at the thought of bringing other engineers into the process, and so the idea of releasing Robertson to the bank did have appeal. He quickly pondered the issue and then granted Robertson's transfer. After a short discussion at Monday morning's joint meeting, Citicorp's management team agreed with the decision and accepted Robertson as its consultant.

◢

On Monday afternoon, Robertson arranged an introductory meeting between the Citicorp officers and two American Red Cross managers specializing in disaster services. These managers, Robert Shea and Mike Reilly, represented an organization with a rich history of emergency aid and assistance.

Founded in 1881 by Clara Barton, the "Angel of the Battlefield," as she became known during the American Civil War and the Franco-Prussian War, the Red Cross was originally conceived to bring aid and comfort to victims of war, but through the early decades of its formal existence, the organization devoted itself primarily to disaster relief.[19] Barton's ceaseless humanitarian endeavors eased the suffering of the injured and raised the spirits of the desolate through a multitude of crises large and small at home and across the world.

In the face of the devastating "Thumb" Forest Fire in Lapeer County, Michigan, in 1881, Barton galvanized the community, raising money and collecting clothing and supplies for the displaced citizenry. In 1889, in the aftermath of the catastrophic South Fork Dam break in Johnstown, Pennsylvania, which killed over two thousand people, Barton coordinated fifty volunteers to provide shelter, aid, and relief to the injured and homeless. When famine struck in Russia in 1892, she arranged for the delivery of six railroad boxcars filled with food. In 1893, when five thousand people died in a coastal storm and tidal wave that decimated the islands of South Carolina, Barton and her organization worked for nearly a year to bring the largely African American population back to physical and economic health. In the aftermath of the Ottoman massacre of nearly two hundred thousand Armenians in provinces of Turkey in the mid-1890s, Barton brought aid and relief to the survivors—the only Red Cross member, and the only woman, permitted access by the Turkish government. And in her final official relief operation on behalf of the American Red Cross, Barton collected and delivered over $120,000 in aid and supplies for the beleaguered survivors of the massive Galveston, Texas, hurricane in the summer of 1900, which claimed over six thousand lives.[20]

Clara Barton's leadership and charitable spirit created an indelible legacy of goodwill and humanitarianism that survives to this day. From managing housing and evacuation efforts during the Great Mississippi River Flood of 1927 to delivering food, shelter, and mental health services following the 1994 earthquake in Northridge, California, to orchestrating the organization's single largest emergency relief effort in the aftermath of Hurricane Katrina in 2005, the American Red Cross was and continues to be the stalwart protector, provider, and healer during times of national disaster. The involvement of the Red Cross during the Citicorp engineering crisis was simply the latest chapter in the organization's long legacy of public service and benevolence.

At the bank's meeting with Robert Shea and Mike Reilly on Monday evening, Robertson revealed the potential of a catastrophic collapse of the Citicorp Tower and the structural problem that prompted it. He explained that provisional measures to monitor the building were under way but that he and the bank's upper management had decided to develop an emergency evacuation plan in the interest of prudence and caution. If the worst should happen, Robertson insisted, a disastrous seriatim failure—one building toppling into the next like a chain of dominos—could theoretically occur.

Shea and Reilly immediately understood the "critical issue" and "dangerous situation" they were facing.[21] They explained that the role of the Red Cross in such a disaster would be to handle the humanitarian side of emergency planning, such as food, shelter, and medical needs. In preparation, they suggested a collaborative effort with a variety of New York City government agencies to develop and implement a detailed evacuation plan for and with the bank. Robertson would later advise representatives of New York Mayor Ed Koch's office of the pending curative measures to the building and the emergency preparedness measures under way.

The men decided that Citicorp management would work independently with the Red Cross and all public agencies during the emergency and that Hal DeFord would become the bank's point person in the development of the evacuation plan. In the series of meetings and conferences that followed, however, it became increasingly clear that the ongoing stress of the crisis was unnerving him. "DeFord was a wreck," recalled Reilly. "It really wore on him. This guy was constantly nervous and the anxiety was heavy on him. He was really going through it. We'd say God . . . wonder if he's going to make it."[22]

14

"WHITE LIES"

TUESDAY, AUGUST 8

President Jimmy Carter was in characteristic good humor on Air Force One on route to New York, and the mood would endure through his entire visit. In town for "happy business," as the *New York Times* wrote, Carter brought with him the fulfillment of a long-term commitment for "serious, responsible, adequate and cooperative" assistance to New York's chronic fiscal problems.[1] Descending the granite steps of City Hall with Mayor Koch and Governor Carey at his side, Carter paused at the podium while the band played "Happy Days Are Here Again." Bearing his wide signature smile, the president told the crowd of five thousand, "This is a good day for New York."[2]

The New York City Loan Guarantee Act of 1978 that Carter had come to sign represented what he called "a crucial step in New York's long and difficult climb back to its solvency and independence."[3] With the goal of stabilizing the city's financial woes and preventing the widespread economic repercussions that would surely result from a default of the nation's largest metropolitan center, the act placed the full faith and credit of the federal government behind the city's debt and provided it with ready access to ongoing credit. "Those who thought that the United States was going to stand by while its greatest city went under were wrong," said the president.

"Let there be no mistake about what this bill does. It is not a handout. New York has asked for no handout and has received none. Nor is it a Band-Aid or temporary approach that simply postpones an inevitable problem. Instead, through long-term guarantees, the bill opens up enough breathing space for New Yorkers to complete the difficult task of restoring yourselves to financial and economic self-sufficiency."[4]

Before President Carter even began his remarks, Mayor Koch handed him an envelope containing a note that read simply, "August the eighth, 1978. Mr. President, New York loves you."[5]

Though the act was significant in its offer of federal intervention to avert the financial collapse of a major US city, it also sparked debate and controversy surrounding federal involvement in local fiscal matters and the potential moral hazard of bailing out financially troubled municipalities. The staunchly conservative Walter Wriston no doubt bristled over the measure, though he was forced to admit its necessity.

As President Carter delivered his lofty remarks at City Hall, Wriston and his management team placed the finishing touches on the language of a press release about the Citicorp Tower's woes. He had delayed issuing a statement for as long as possible, but it had now become clear that without some plausible and calming public explanation for the brigade of torch-bearing welders that would soon be ascending this major metropolitan skyscraper, a tabloid maelstrom would be inescapable.

Throughout the crisis, LeMessurier had communicated regularly with Alan Davenport and Nick Isyumov about weather projections and to fine-tune the wind tunnel data as it related to the Citicorp Center vicinity. When LeMessurier flew up to Canada at the end of July to meet with them upon first learning of the tower's sensitivity to quartering winds, one of Davenport's associates, David Surrey, told him that based on recent analysis for another proposed building in New York City, estimated wind velocities in the area were actually statistically *higher* than were thought at the time of the Citicorp Tower's original design.

LeMessurier had been confused by the data, but now, as the prospect of a public statement on the repair project loomed, he recognized this new information as a godsend. Perhaps, he thought, Surrey's revised meteorological data could be used by the bank as a pretext for the required repairs without having to acknowledge the underdesigned connection bolts or the bracing system's inherent vulnerability to quartering winds. "I clung to that statement," LeMessurier recalled.[6]

The premise, as he suggested to Citicorp's management, was that once armed with that "little piece of knowledge," he reached out to the bank and offered to upgrade the building and to improve its longevity as a simple matter of prudence at a fairly modest expense.[7] "I actually had new information, that's true," he later said. "So this was a softening. Doesn't mean we were covering up and hiding, in fact we were taking all the action it was necessary to take, but there was no necessity of letting people know the sense of horror on our minds. Was there?"[8] Robertson had a somewhat different take on the matter. "This remark by Mr. Surrey, later repeated to high executives in Citicorp by Mr. LeMessurier, seems to have caused considerable confusion and reduced the general understanding of the seriousness of the problem," he later wrote.[9]

LeMessurier had formalized his proposal to Citicorp in a written statement intended as a safe harbor for the bank's reliance rather than for actual public release. It read,

> During the design stage for Citicorp Center, we conducted extensive wind tunnel tests at the University of Western Ontario. This wind tunnel is the most sophisticated and experienced in the world. A building model was tested to determine both static and dynamic forces. These tests for Citicorp were conducted in 1973.
>
> Within the past months, we learned that recent studies for another proposed building in midtown New York, also tested at the University of Western Ontario, show an increase in maximum probable wind velocities of 9%. The effect of this change is to increase maximum possible wind forces on Citicorp Center by up to 25%.

The bank, with my advice as its engineer, has decided that . . . it would be prudent to increase the performance characteristics of the wind bracing system to reflect this increase in expected forces. The maximum resulting strength and stability will substantially exceed the standard required by the New York City Building Code.

The work consists primarily of reinforcement of joints in tension and is expected to be completed in three months.[10]

"We had to cook up a line of bull, and white lies at this point are entirely moral," LeMessurier later confessed. You don't want to spread terror in the community to people that don't need to be terrorized."[11]

During breakfast at the Waldorf on Tuesday morning, LeMessurier and Robertson recognized that "good politics" would dictate the order of the day's events. They knew that Citicorp was finalizing its public statement and was about to release it to the media. Certainly the bank could not inform the press of the crisis before notifying local government agencies, and so Robertson had hurried together a meeting with the city for the same day. "You don't want public officials to hear about it secondhand," LeMessurier later said. "They'll hold it against you forever."[12]

Later that afternoon at Robertson's Park Avenue office, LeMessurier candidly revealed the Citicorp Tower structural flaw and the possible danger of collapse to several members of the New York City Building Department, the chief engineer of the city planning commission, and a representative of Mayor Koch's office. He held the floor for over an hour and openly confessed the lapse of his office in fully appreciating the significance of quartering winds and the increased forces they posed on the Citicorp Tower's bracing system as constructed. He explained how he had discovered the vulnerability and provided a detailed proposal for its remedy. "I told the whole story, everything," he remembered.[13]

The tone of the meeting was cordial and focused. To a man, the city officials were grateful for LeMessurier's candor and supportive of his proposal for corrective action. To his relief, there was no demand for any protracted review process or desire for detailed inspection or oversight of the work. The absolute absence of administrative meddling in a room full of government officers and bureaucrats was, to LeMessurier, as welcome as it was surprising. After a brief discussion and a few questions, the men requested only that they be kept advised of the project details through the filing of all plans and drawings and that the work be completed in accordance with code. LeMessurier later noted, "The opinion was openly expressed that the engineers had behaved in a very professional manner and that the City intended to help in any way to expedite the work."[14] He recalled one of the officials jesting that if everyone acted as he had, there would be no need for the building department.

Then, DeFord nervously informed the gathering that the bank was in the process of releasing a statement to the press about the matter that may not be as forthcoming as it could be. One official, Lee Oberst, a former National Guard artillery officer and efficiency expert who worked in both the Beame and Koch mayoral offices, drolly responded, "If I'm asked the question of what will happen to this building if there's a big wind before it gets fixed, I'll say there will be 'evident signs of distress.'" The room erupted with laughter. "Those words I remembered very well," LeMessurier later said.[15]

Still, something troubled Les Robertson. He had remained uncharacteristically silent throughout the meeting. He noticed that LeMessurier, in his presentation, had explicitly informed the city officials that the structural system of the Citicorp Tower as originally designed and constructed adhered entirely to the requirements of the New York City Building Code. At the time, Robertson had already begun a comprehensive design review of the entire building, and he knew that LeMessurier's statement was not totally true. Though Robertson flinched at the assertion, he did not challenge it. He later explained, "Armed, perhaps, with more knowledge of conditions to be found in the building than was Mr. LeMessurier, but respectful of his position as the engineer who carried primary responsibility for the design conception of the structure, we did not dispute his statement of conformance with the Code."[16]

That professional deference would soon be tested.

◢

The on-site MTS technicians had requested daily notifications of weather changes in order to execute real-time adjustments to the TMD in accordance with current and expected wind conditions. LeMessurier's associate Al Romaneski, who remained in New York for the duration of the crisis, generated a procedure with Robertson to relay to MTS daily and long-term storm forecasts obtained from Irving Singer and the National Weather Corporation. Each morning, a representative from MTS would contact Robertson's office to retrieve the data, and through the exchange of telephone numbers, the company could be advised at any moment of sudden weather changes, should they occur.

LeMessurier had placed virtually all of his confidence in the building's resilience on the full functionality of the TMD.

◢

FOR IMMEDIATE RELEASE:

NEW YORK, August 8—Citibank said today that a review of Citicorp Center's design specifications was recently made by the engineers who designed the building in connection with their current designing of another building of similar characteristics.

This review has caused the engineers to recommend that certain of the connections in Citicorp Center's wind bracing system be strengthened through additional welding. Work has already started and will be completed as soon as possible. The engineers have assured us that there is no danger.

The City Government has been notified and told us that the Building Department will give full cooperation.[17]

Even before the issuance of Citicorp's formal press release, the gathering of New York City officials at Robertson's office had piqued the interest of local newspaper reporters. "Our telephones were going off like crazy. . . . Took very little time for [the press] to figure out that in our office was just about everybody in town."[18] Robertson, however, did not wish to be involved in a political maelstrom and shied away from media attention. He allowed others to respond to the press inquiries and largely evaded the public eye during the crisis. "I have no interest in such matters, and Citicorp had very competent people to worry about such things," he said.[19]

Now Lamson Smith's phone began to ring. As vice president in charge of Citicorp's Public Affairs Department, Smith had issued the statement on behalf of the bank but had opted to refer follow-up press inquiries directly to Hal DeFord.

"We're a very cautious organization—we wear both the belts and suspenders here," DeFord told reporters. "The building was designed in 1971 on data then available. As it is, the building could sustain a '100 year wind' a wind in excess of 93 miles per hour for a period of 10 to 20 minutes."[20] After the fix, he insisted, the tower could withstand 125-mile-per-hour sustained winds—well beyond the highest recorded wind limits in Manhattan. "We don't want people concerned, so we sent out a press release announcing the work," he said casually.[21]

Though the bank's statement had been purposely crafted to impart an innocuous, matter-of-fact tone, it was met with immediate skepticism. Questions about the building's safety began to circulate, and Citicorp's effort to play down the story seemed only to heighten the interest. A reporter from the *New York Daily News* telephoned the acting building commissioner, Blaise Parascandola, who had attended the meeting at Robertson's office and been directed by Mayor Koch to investigate reports of danger. "Parascandola seemed satisfied with the proposed solution," wrote the paper. "Of course it's improbable," he said, "but there's always a chance of winds up to 150 mph, which, at certain angles, could break . . . and crack walls and glass, and put stress on the basic structure of the building. This way we'll be safe."[22]

Soon after, another reporter—this one from the *Boston Globe*—called LeMessurier and asked him directly to respond to the rumors. By then, the TMD had been fine-tuned, and the engineer was confident in its ability to steady the building in the event of extreme weather. "Not so," he said when the reporter claimed that the tower might be in danger.[23]

⊿

But LeMessurier's pretense for the Citicorp press release proved to be just that. At about the time the bank issued its statement— and, perhaps, before—LeMessurier received a telephone call from Nick Isyumov at the University of Ontario Wind Tunnel Lab. The "new information" he had received from David Surrey in late July claiming an increase in statistical wind forces in New York—the information on which the bank had based its public statement— had been incorrect. In fact, Isyumov confessed, the data indicated a favorable *decrease* in velocities.

LeMessurier knew that the bank's press release was already in progress but still believed it to be technically accurate. If the entire truth were to be made public, he rationalized, it could potentially create unnecessary panic and damage to his and other professional reputations. While the information was undeniably good news, he was puzzled by the data and needed to learn more to understand its full import. Isyumov informed him that Alan Davenport's mother had recently died at her home in England and that Davenport was there tending to arrangements. After the funeral, Isyumov promised, they would both come to New York and present their full and most current data.

⊿

"Stone cold lies," said Jim Dwyer, a writer for the *Daily News*, of the bank's press release. He called DeFord and asked him to comment on reports that the Red Cross had even been mobilized in the area to respond to the danger. "I don't know of any involvement whatsoever with the Red Cross," DeFord reportedly said. When Dwyer pressed him about talk of evacuation planning, according to Dwyer, DeFord responded, "Doesn't ring a bell with me."[24]

15

"A THOUSAND-YEAR WIND"

WEDNESDAY, AUGUST 9

No sooner had Tropical Storm Bess changed direction and lost its punch over the Gulf of Mexico than Robertson's weather forecasters began monitoring another disturbance churning off the northwest coast of Africa. Initially designated as a tropical depression, the storm quickly intensified and, upon the emergence of a distinct eye, was upgraded by the National Hurricane Center. A meteorological rarity, Hurricane Cora became only the second storm in recorded history to gain hurricane status solely through satellite observations.[1]

Amid consensus forecasts of continued strengthening, Cora spiraled into the southeastern Caribbean Sea, only to unexpectedly lose circulation and, much to Robertson's relief, dissipate as quickly as it intensified. The Winward Islands would escape damage from the storm, but throughout the day of August 9, officials in Dominica, Martinique, Saint Lucia, Saint Vincent, Grenadines, and Grenada held their collective breath as Cora approached and peaked.

John Woolley and Irving Singer would continue their scrutiny of the seas, aware that any detected disturbance could intensify and career northward toward New York City.

LeMessurier's "line of bull" found its way into Wednesday's newspapers in New York and beyond. The *New York Daily News* carried the story under the banner, "Citicorp Bldg. to Get 1M Wind Bracing," while the *New York Post* claimed, "Citicorp Tower Getting 'Braces.'" The *Boston Globe*, which had picked up the story from United Press International, the worldwide news wire, warned, "Citicorp to Brace for Big Wind," and the *Wall Street Journal* went with the headline, "Citicorp Tower Gets More Steel Bracing as Added Precaution."

Before long, LeMessurier was dodging interview requests from multiple media outlets, but he had learned from Walter Wriston that evasion breeds doubt. On the calls he did accept, he remained judicious in the information provided, speaking only in broad strokes and sweeping generalities. "Bill LeMessurier of LeMessurier Associates has assured the bank that there is no current danger to employees or pedestrians," wrote the *Boston Globe* after confirming the comment with the source.[2]

Earlier in the year, LeMessurier had granted an interview to a young journalist from *Engineering News-Record* named Jane Rippeteau for a story about an ongoing project at King Kalid Military City in Saudi Arabia. Upon learning of the Citicorp story, Rippeteau called and asked, "Bill, what's going on there, can you tell me about it?" He trusted and respected the reporter and did not want to arouse suspicions by seeming evasive about the subject. "I suppose I can, why don't we have breakfast at the Waldorf," he responded.[3]

The interview was cordial but probing. While LeMessurier did share some technical details of the contemplated structural repairs for the benefit of *Engineering News-Record*'s target audience in the design industry, for the most part he stuck to his narrative that the improvements were governed by prudence rather than necessity. "I advised the bank and they listened to me," he told Rippeteau when she asked whether the repairs were indeed optional.[4] "I told her half-truths," he later acknowledged.[5] Near the end of the article, titled "Engineer's Afterthought Sets Welders to

Work Bracing Tower," the reporter quoted Hal DeFord as saying that Citicorp was currently paying for the corrective work but hoped to be reimbursed through insurance.[6] "I know she smelled a rat," LeMessurier later recalled with a wry grin.[7]

The *Engineering News-Record* article would not be published for another week, but events occurring later that evening would render LeMessurier's greatest media concern, the New York City press, all but irrelevant.

◢

Arthur Nusbaum coordinated the repair work on the Citicorp Tower with welders to be provided by Karl Koch Erecting. Robertson had earlier notified the governor's office of the Citicorp Tower troubles and the plans to cure them. Now, at Robertson's behest, William Spencer requested the state's consent and guidance on the repair work. A representative of the New York Department of Highways, Warren Alexander, was assigned to the project to assist Robertson with the efficiency and reliability of the welding process. The move to seek state oversight on the project was, undoubtedly, intended to mitigate the potential for any antagonism, since New York City officials had already been advised of the operation and consented to it. Alexander reviewed the remedial plans, offered several suggestions, and approved implementation of the work.

With the hurricane season well under way, the bank's management team had assumed that welding would be conducted around the clock. However, Nusbaum knew the process would produce a noxious haze that could both panic and physically discomfort the building's occupants. To minimize any unpleasant effects and to conceal the project as best they could, the men agreed that the work would be conducted only after business hours but throughout each night, seven days a week.

◢

Through much of August, LeMessurier's wife, Dorothy, had remained with her husband at the Sinatra Suite at the Waldorf in

New York. She had been worried about the emotional and physical toll the situation was taking on her husband, and she thought it best to be by his side. At the very beginning of the crisis, the couple had stayed for several nights at the St. Regis Hotel on East Fifty-Fifth Street and Fifth Avenue, but when the Waldorf suite became available, it would conveniently serve as both LeMessurier's office and their temporary residence. As the work intensified and the suite filled with bankers, engineers, contractors, and consultants, Dorothy became not only a protector to her husband but an administrative assistant running errands, preparing drinks, and managing the phones for the crisis center.

LeMessurier's day had been particularly demanding. Weather reports of disturbances in the Caribbean continued to stream into Robertson's office; any one of those, LeMessurier knew, could trigger the pager alarm on his belt. The repair work on the tower was only just beginning, and "the unpredictable" still lurked. Robert Dexter had requested LeMessurier to provide a formal letter itemizing when and in what manner he had first communicated word of the tower's structural problem to the bank. It was, no doubt, an initial legal step and a possible sign of things to come. LeMessurier had greeted Alan Davenport and Nick Isyumov, who had flown in from London after the funeral of Davenport's mother. There remained the issue of conflicting New York City weather data provided by the wind-tunnel experts and whether the Citicorp press release had, in fact, been based on faulty information. And there was the horde of newspaper reporters that LeMessurier had dodged throughout the afternoon. So it was with some trepidation that Dorothy informed her husband when he returned to the lavish confines of the Waldorf suite at the end of the day that a reporter for the *New York Times* had called and wanted to speak with him.

LeMessurier instantly tensed at the news. He knew that the *Times* was an international media gateway, and if it carried a story about the Citicorp Tower structural problems, any thought of somehow containing the damage—or salvaging his career—was over. He telephoned Carl Sapers and asked his advice, secretly hoping the lawyer would instruct him to ignore the request.

"They'll smell a rat," said Sapers, if he did not call back.[8] LeMessurier sighed and thanked him for his time. He fixed himself a martini and glanced at the clock. It was just after 6 p.m. He picked up the room phone and called the number provided by the reporter, uncertain of exactly what to say or how to respond to the questions that would surely come. After several rings, the call answered with a recording.

As of 6 p.m., the *New York Times*—and the other prominent metropolitan newspapers—had gone on strike and suspended operations.

THURSDAY, AUGUST 10

Hugh Stubbins later reflected, "If the newspapers hadn't been on strike, I dread to think what would have happened. It would have been pure hell."[9] It was, said LeMessurier, "the greatest thing that ever happened."[10]

The New York Pressmen's Union work stoppage of 1978 had little to do with the traditional labor calls for increased wages or improved working conditions. The action was the simple result of, and anger toward, the march of technology. The automated machinery of newspaper production had, in essence, rendered a portion of the industry's labor force obsolete, and management wished to "trim the fat." "So long as the presses are running smoothly, with no breaks in the web of paper or problems of inking and folding, pressmen necessarily must be idle for a substantial portion of their shift," reported the *Times*.[11] The expired contracts, which were the subject of intense renegotiations, required management to hire up to twice as many pressroom workers as they actually needed. The requirement, insisted publishers, placed them at a competitive disadvantage with their suburban peers.[12] Unable to reach agreement on the issues of overtime, availability of shifts, additional hiring, and attrition of workers, the union posted picket lines on the evening of August 9 and railed against the "greedy for profit" owners.[13] Believing, perhaps, that the strike would be short-lived or that the newspapers could continue to publish without union labor as the *Washington Post* had done in 1975, ownership persisted in its restructuring effort and refused the union's demands.

The *Times* and the *Daily News* would remain shuttered for the next eighty-eight days, while Rupert Murdoch's *New York Post* resolved its labor issues in fifty-six days, upon agreeing to accept whatever the other papers would ultimately negotiate with the striking workers. Though, in the end, the publishers gained considerable concessions from the union, they would be forced to admit some "major misjudgments." The combined losses in advertising revenue to the three newspapers in the wake of the strike would approach $150 million, while alternative media outlets in New York—television, radio, magazines, and nonunion newspapers—enjoyed an explosion of patronage and sponsorship. New York's power-house newspapers were left with an uphill battle to recover their lost readership.[14]

During the newspaper strike, the New York Yankees, who had floundered in the early part of the 1978 season with clubhouse drama and toxic management, surged to a World Series victory, free from the distraction of venomous sportswriters. "The strike, coming when it did," said Bob Lemon, who had replaced the volatile Billy Martin as Yankee manager midseason, "did more for us than if we picked up a 20-game winner."[15]

William LeMessurier, it seemed, was not the only beneficiary of the pressmen's strike.

On Thursday morning, LeMessurier met Les Robertson and a welding expert, Jim White, Robertson's longtime associate who was also a part of the World Trade Center engineering team, to discuss welding protocols at Citicorp Tower. They were joined by Warren Alexander, the state's welding consultant. The order and priority of structural repairs, they agreed, must be governed by accurate wind and climate projections.

Davenport and Isyumov conferred with LeMessurier upon their arrival in New York, to address the confusion regarding wind projections in New York City. They confirmed that while the revised data relied on by the bank in issuing its press release may have been inaccurate, LeMessurier's initial fear that the Citicorp Tower was in grave danger was nonetheless verified.

At a conference at Robertson's office that afternoon, the wind experts presented their findings to DeFord and Dexter as well as Hugh Stubbins, Warren Alexander, and the structural engineers. Based on their analysis of the latest historical weather data projections for New York City and after updating their extensive wind-tunnel calculations, Davenport and Isyumov told the gathering that they were able to accurately predict, within the state of the art, the expected performance of the tower in a full range of weather conditions—and the news was not good. LeMessurier later summarized the presentation: "Taking into account new wind records, the exact measured dynamic properties of the building and detailed study from all directions the predicted storm that would produce collapse was exactly the same sixteen-year storm . . . first estimated from the data of the original 1973 study."[16]

LeMessurier had already begun detailed planning of the priorities for the corrective welding, and he assured the group that with Davenport and Isyumov's updated wind and weather assessments, he could produce a schematic evaluation of the tower's bracing system, assigning the most vulnerable elements and connections the highest level of importance. "By ranking such condition in terms of rarity of the storm required to produce failure (which correlated with the wind velocity at failure) a table of work priorities [could be] established," he wrote.[17] Through this process, the bank and the engineers would be able to quantify the tower's restoration as work progressed. With each completed weld, the statistical probability of a catastrophic structural failure would correspondingly drop. "You fix this and then the rarity of the storm that will cause any trouble lengthens, with two variables, with and without the damper going," he told them.[18]

That afternoon, Stanley Goldstein, whom LeMessurier had appointed engineer in charge of the remedial work, provided the foremen and deputies of HRH and Karl Koch Erecting the final details of the work to be done and the field procedures to be followed. They agreed that, in the interest of safety and discretion, the welders would reduce preheating for welds to the lowest possible values, and they established protocols for communication and in-field decision-making. The goal, Goldstein implored, was to conduct the work in as judicious and timely a manner as reasonably possible.[19]

◢

Beginning on the warm and windless evening of August 10, 1978, and every subsequent night for the remainder of that summer, a slew of over twenty face-armored union welders loaded Karl Koch's steel plates and equipment through Citicorp's loading docks and ascended service elevators to various floors selected according to LeMessurier's evolving list of priorities—sans the probing gaze of New York's major newspapers.

The procedure was surgical. A crew of union carpenters and drywall contractors assembled a series of plywood enclosures at designated locations, behind which they exposed the steel chevron splices to be repaired, removing sheetrock like a surgeon opening an ailing patient. By morning, when staffers returned to work and opened their offices, the wood enclosures would be gone and the walls repaired. Wriston had insisted that all tenants be made aware that work was to be done in their spaces, but the actual purpose and necessity of that work was kept a guarded secret.

"It was a tense time for the whole month," LeMessurier later admitted. He had shouldered the ongoing process of evaluating and reevaluating the building's vulnerabilities and had synchronized the order of work to be performed. He carried around detailed diagrams and statistical charts and incessantly updated his calculations so as to correctly orchestrate priorities and track progress. "I was constantly staying on top of which joint to fix next, and which level of the building was more critical," he recalled.[20] He had classified over two hundred connections as "critical."[21]

LeMessurier's team of engineers had retrofitted each steel plate to fit precisely across both sides of the diagonal member connections like rigid leg braces on either side of an ailing knee. The existing bolts that spliced together each adjoining segment had been fastened through drill holes at the milled end of each member and, for increased overall strength, were allowed to remain in place as designed. The plates simply upgraded the system's original bolted strength rather than replacing it. In this way, the size and weight of the corrective sheaths and the actual weld volume could be reduced. Robertson noted that the new penetration welds would actually be stronger than the girders they fused and that had the original design of the bracing system not been altered from welded to bolted splices, "the connections would have been stronger than the members themselves—and very substantially stronger than was required."[22]

The engineers had estimated that 216 of the tower's chevron connections would require buttressing with a total of 544 steel plates, each one and one-half inch thick and weighing two to three hundred pounds.[23] According to Robertson, an additional 120 to 150 tons of steel would be added to the Citicorp Tower.[24]

"We decided as long as we are fixing these things, we're going to make the connections so that the building will not fall down in a thousand-year wind," LeMessurier later said.[25]

16

SAFETY, SECRECY, AND SURVEILLANCE

As corrective welding on the tower progressed and its safety slowly improved, the wind-velocity trigger for evacuation of the building and its surrounding areas theoretically began to ease. An informal chain of authority had been created for making the "go/no-go" decision on evacuation, which, of course, began with the meteorological data obtained through Robertson's weather forecasters. From these projections, which were updated two to three times a day, LeMessurier and Robertson were jointly tasked with determining whether to recommend activation of the plan. "We would report to an operations officer [at the bank], and then there would be either an alert yes or an alert no," recalled LeMessurier. "If no, you simply inform. If yes, it would go to a senior executive of the bank. He'd report the emergency and he'd report in two directions—implement plan and talk to director of city operations."[1]

The idea of shared command between LeMessurier and Robertson was, of course, unworkable, since each of them had different levels of confidence in the TMD and thus different thresholds for issuance of an evacuation order. LeMessurier believed that the device could be relied on to increase the tower's wind resilience in all but the most extreme weather, while Robertson favored a more conservative approach to its reliability. Both men equally understood the vulnerabilities of the building, and each shared the ear of the bank's executive team. Citicorp, in conjunction with local government officials, would make the final decision on whether to evacuate after evaluating the advice of both engineers, even if their advice differed. LeMessurier privately believed his opinions would carry sway with the bank in the event of a conflict since he alone had been tracking the progressive safety state of the building with detailed computations and methodical charts, and he was convinced of his ability to predict the tower's response to a full range of conditions. "I think the bank trusted me," he said in retrospect.[2]

The evacuation plan itself had been shrouded in secrecy from its conception. In early August, Mike Reilly of the American Red Cross had attended a briefing of the mayor's and governor's offices and recalled the meeting to be an "ominous event, pretty serious business."[3] That evening, the respective principals from both government offices personally advised Mayor Koch and Governor Carey of the Citicorp Tower structural problem and the planned emergency measures.

From the start, Reilly had been troubled by the lack of public disclosure in the formulation of the Citicorp evacuation plan. "Most of what we would do would be working hand in hand with government [and] with other agencies. It would be full public information," he recalled. "It was very clear right from the start that this was to be done confidentially, that it was being handled on the highest levels of industry and on the highest levels of government, and that our work was to be as confidential as possible."[4]

The initial focus of the emergency planners was to determine the number of people in and around the vicinity of the tower so that in the event of a high-level storm threat, the Red Cross and city officials would know how many people needed to be removed to safety—or in the event of a catastrophe, how many souls would need to be accounted for.

According to Reilly, Citicorp and the Red Cross recruited a team of parishioners from a local church attended by one of the bank's officers to conduct a house-by-house, business-by-business canvass within a designated perimeter surrounding the building. The work was conducted under the guise of a "marketing survey" and with as few actual Red Cross personnel as possible so as not to arouse concern or suspicion. There was no doubt, however, that should an evacuation become necessary, thousands—perhaps tens of thousands—of people would be affected. As LeMessurier later elaborated, "The plan was to get the police moving and get in touch with everybody in every building within ten blocks. You take a drawing of New York and where would the building fall? It would fall in any direction, it might fall on Bloomingdale's. . . . But there was no question from the very beginning, that if the building collapsed it would hit something. . . . It would be a major urban catastrophe and there would be people at risk who had nothing to do with the building."[5]

Even as welding work on the tower continued, LeMessurier remained cautious in his assessment of the building's safety. He had determined that the bolted joints on the thirtieth floor—at about half the height of the structure—were the most susceptible to quartering winds and, thus, should be urgently corrected. A "peculiarity" of the building, as he called it, made it extremely sensitive to the failure of a single joint, and so he feared that a wind with the power to tear the most vulnerable seam apart would probably lead to the progressive failure of the entire structure.[6]

Yet LeMessurier constantly balanced those concerns against his dread of an unwarranted evacuation order. He had always derided Robertson's unwillingness to rely on the TMD, even as modified, as overly cautious, and he continued to believe in its effectiveness. The device was tested multiple times each day by the manufacturer and monitored around the clock during the crisis. Fully operable, the damper enhanced the building's durability and significantly decreased the probability of failure during extreme weather. But so long as "the unpredictable" was still possible, LeMessurier knew that the tower remained in danger.

One night in mid-August, LeMessurier was jarred from sleep by the shrill alert of his pager. The warning system, instituted by Kinemetrics as an adjunct to its surveillance of the building, was designed to activate when and if the sensing instruments throughout the tower registered dangerous stress levels or unusual sway. The pager alert was never going to mean good news.

He quickly called Robertson, whose pager had also sounded, but neither man knew what triggered the alarm. The winds in Manhattan had been calm, and the forecasters had not issued any warnings. Robertson immediately telephoned the on-call Kinemetrics representative, who was already anxiously scrutinizing the electronic monitors that fluttered and danced and then inexplicably went dead. Was there a problem with the building? Had the underdesigned splice bolts finally given way?

As LeMessurier and Robertson breathlessly awaited word, the Kinemetrics rep scouted the Citicorp Tower. The weather was obviously not a factor, and he could detect no structural movement or disturbance. As he began inspecting the company's intricate electrical wiring, however, he realized that many of the wires woven through the system had been severed. Perplexed, he called Robertson and explained what he had found. Robertson, having begun his career as an electrical engineer and having spent more than a decade in New York, immediately recognized the source of the problem. He asked whether Kinemetrics, a California company had, perchance, hired New York union electricians to wire the system or whether they had used their own nonunion company personnel to do the work. He already knew the answer.

Later recalling the event with some amusement, LeMessurier explained, "Well, these clowns came in from California, and what they didn't know is you do not do wiring in New York except with union electricians. The union heard about it and responded appropriately."[7]

As quickly as Kinemetrics repaired the wires, they were again—and repeatedly—cut in the dead of night by disgruntled union members in what Robertson later called "a continuing effort to sabotage the system."[8]

Even when the tower's stress monitors were fully functioning, LeMessurier questioned their value. While he professed confidence in the system to accurately monitor the strain levels in the tower, he feared that by the time acute stresses were detected, there would be little opportunity for response. "If they had gone off the chart it would have been too late to evacuate the building. You had to make that decision on the basis of forecasts," he later said.[9] Robertson was more optimistic in assessing the instrumentation installed by his California friends. "These data . . . would allow at least a few moments warning in case of a dangerous level of response of the building to wind-induced excitation," he later wrote.[10]

It was with much embarrassment that the engineers reported the initial incident concerning the building monitoring system to the bank's officers. LeMessurier was adamant that he and Robertson be allowed open and productive lines of communication with the highest levels of Citicorp's executive team. William Spencer, the bank's president in charge of the repair project, agreed. The building alerts, he realized, benign as they turned out to be, served as an example of the need for constant interaction between the principal parties. He encouraged continuous meetings to discuss the status of the tower's recovery, and he implored the engineers to report any bank official—including himself—who failed to properly do his job directly to the bank's board of directors. Every Wednesday night during the crisis, Robertson met with Spencer and Chairman Wriston to chat about the status of the project and to address any problems that had been encountered. Bureaucracy and corporate backbiting was not going to derail the urgent work ahead.

After organizing the priorities and procedures of the initial welding process and corroborating on a workable course of communication between the contractors, engineers, and the bank, LeMessurier finally decided he needed a break. The weather in New York had remained mercifully tranquil, and so, on the morning of Saturday, August 12, he and Dorothy flew to Portland,

Maine, and made their way to their island retreat on Sebago Lake. The respite from the ongoing drama in New York, however, lasted mere hours. LeMessurier was informed that DeFord and Dexter were planning a meeting at the bank on Monday to refine the evacuation criteria and that his presence was required to provide up-to-date data on the welds and the health status of the building.

LeMessurier had been worried that perhaps the officers of Sippican Consultants International, the organization that had purchased his company in 1973, were angered that he had not spoken to them before taking the action that he did with Citicorp. When the company offered to send its private jet to Portland to shuttle LeMessurier and his wife back to New York on Sunday night, it was met with both gratitude and relief.

As the plane began its final approach to LaGuardia Airport, LeMessurier peered out across the East River to Manhattan and immediately spotted the Citicorp Tower "lit up like a Christmas tree," from the spotlights and welding torches of the union workmen. It was, in the words of one writer, "a pillar of fire on the Manhattan skyline." Thankful for the ongoing pressmen's strike in the city, he said to Dorothy, "Isn't this wonderful? Nobody knows what's going on, but we know and we can see it right there in the sky."[11]

17

A CRITICAL ANALYSIS

Even before the shifting of Robertson's professional responsibility from LeMessurier to the bank, he had begun a formal and thorough evaluation of the Citicorp Tower's full structural system beyond just the bracing members vulnerable to quartering winds. Trained on the West Coast in earthquake design and resistance, Robertson was, perhaps, more attuned to the importance of load paths in high-rise bracing systems than were his East Coast counterparts.[1] When he first reviewed the building's design drawings during his initial visit to LeMessurier's New York office on August 2, he had identified what he believed to be a variety of technical deficiencies with the structural regime of the tower. Sufficiently alarmed by his discoveries, he insisted that an immediate and comprehensive review of the building's entire support system be undertaken not only to identify any structural flaws beyond the bolted chevron connections but also to assure the bank—and ultimately the public—that *every* element of the tower's inner structure was either properly designed or adequately corrected.

LeMessurier and Robertson agreed that three engineers from LeMessurier's Cambridge office who had no prior affiliation with the Citicorp Building would be named to a design review team and stationed at Robertson's Park Avenue headquarters. The concern, of course, was whether the team members could conduct their work objectively, even if the results turned out to be unfavorable to their primary employer. But Robertson's existing office staff was already overworked, and he wanted the financial burden of the effort be placed on LeMessurier, the original designer of the structure, so he agreed to use the three engineers for the review.[2]

Within hours of the decision, an officer of Sippican Consultants instructed Richard Henige, one of the company's young engineering associates, who had just returned from his honeymoon, and two of its seasoned project managers, Rolf Andersson and Wayne King, to drop everything and immediately travel to New York for an extended stay. They were given no further explanation. The next morning, LeMessurier personally met the three perplexed men at the airport, briefed them on the Citicorp Tower structural problem, and told them they would be working on an independent and highly confidential forensic assessment of the building under Les Robertson's supervision.

As the welding process began and through the ensuing months, Henige, Andersson, and King worked tirelessly on the analysis of the building—from the stability of its legs to the wind-bracing capacity of its crown. Completely discounting the original design computations performed by the LeMessurier-Ruderman venture in the initial planning of the building, the three engineers essentially started from scratch, creating brand-new calculations for every major component essential to the structural integrity of the building. As Al Romaneski later wrote, "The effort was quite prodigious, time-consuming and thorough."[3]

Working long and intensive hours and aware of the urgency of the task, the men developed a three-dimensional, computer-generated, mathematical model to test each member of the building under various hypothetical stresses, and they carefully analyzed the structure's primary frame for wind loading and overall stability. They quickly learned that the deficiencies in LeMessurier's structural design were even greater than previously feared.

For LeMessurier, it was a time of agonizing personal strain. Though he dared not quarrel with Robertson's insistence or the team's work, it seemed that his structural design—even his aptitude as an engineer—was being challenged by every detected and inevitable flaw in the building. In essence, Robertson had demanded a schematic deconstruction of the building—what one commentator called a "post-construction autopsy."[4] The results, LeMessurier knew, would either absolve or condemn him.

First, the team confirmed what LeMessurier and Robertson already knew: the entire Citicorp wind-bracing system as constructed was vulnerable to nonperpendicular wind forces. Although the diagonal members of the system were themselves adequately designed for strength, the engineers agreed that failure of one or more of the bolted connections holding those members together would probably result in the collapse of the entire structure. Hence, they endorsed the welding repairs already in progress.

In the course of the team's work, however, they discovered that the problem went significantly beyond the bolted connections.[5] The audit revealed that a variety of horizontal ties, lateral braces, and truss members at various locations in the tower were also susceptible to quartering winds and required strengthening. Furthermore, a series of horizontal beams that contributed to the overall stiffness and stability of the building—the so-called chevron spandrel girders—were found to be substantially underdesigned and, thus, overloaded by nonperpendicular wind forces. The team recommended repair and bracing of those spandrel members at various points.

When Robertson first reviewed the Citicorp design drawings, he quickly noticed a troubling lack of bracing strength in some compression members located within the central core of the building between the elevator and ventilation shafts. "When these members tend to buckle into the building," Robertson cautioned, "they cannot do so without crushing the floor slab. It is the need to hold columns and main diagonals onto the building that is the issue—for, if there is not adequate tensile capacity, the member will simply tear away from the floor slab."[6]

The review team noted that the New York City Code as well as sound engineering practice required that these elements be adequately braced to withstand certain force levels. In the case of the Citicorp Tower as built, however, it was virtually impossible to attain those strength levels short of reconstructing the entire building. For this reason, and since the code requirements were somewhat clouded, Robertson employed, at LeMessurier's expense, Joseph Yura from the University of Texas, at the time one of the country's leading authorities on steel-structure stability and connection strength.

After a detailed analysis of what was termed the "wandering column" problem, Yura, in collaboration with Robertson and the design review members, arrived at acceptable criteria that comported with practical reality, though not necessarily code. "Larger values were simply not possible to achieve," wrote Robertson.[7] As a result of this study, the team recommended and Robertson agreed that lateral support bracing should be added between the four mast columns as well as two interior core columns and the corresponding floor beams at 104 locations above the fourteenth floor of the tower to add strength and stiffness to these members.[8] LeMessurier privately fumed at the requirement. "The four exterior [mast] columns had floor beams intersecting them at each floor. Such beams would be considered adequate bracing by most engineers," he wrote.[9]

A further concern to Robertson and the design review engineers was whether the tower floors had been properly engineered to also function as rigid diaphragms for the distribution of wind loads between the inner core and the wind-bracing system. In most buildings, concrete floor slabs provide ample strength and capacity for this dual role. In the Citicorp Tower, however, the floors were composed of corrugated metal or cellular decking gapped twice on each level for ducts that carried electrical and communications cables. So-called trench headers that ran wall to wall severing each of the floor diaphragms were utilized to gather these cables but were not filled with concrete, and thus, all wind and compression forces were transmitted through un-reinforced thin steel plates located at these headers. LeMessurier noted that "the strength analysis of these floors was extremely complex. No tests of the electrical trenches were known, and Robertson believed these trenches to be a serious weakness." It was, LeMessurier acknowledged, "a most intractable and unpredictable problem."[10]

The design review engineers performed some hand calculations to determine the strength of the trench headers but, without the benefit of actual testing, were not confident in their results. They searched for any evidence that the original engineers had independently established the strength of the floor system but found none.

Again Robertson strongly recommended an expert. At LeMessurier's request, Clarkson "Pinky" Pinkham, a West Coast specialist in metal deck flooring systems, rushed to New York to evaluate the Citicorp floor design. As the review team had found, in the absence of actual site testing, Pinky could conclude only that the strength of the floor diaphragm where the trench headers bisected the slab was inadequate. Without uncovering an actual site where the apparent weakness could be analyzed, there was simply no way to know for sure.

With all the talk of inadequate bracing, buckling floors, and wandering core columns, the bank officers were understandably getting frustrated. Hal DeFord in particular was angered by the ever expanding scope of the project and the additional time and expense that were required for the remedial work. Robertson and LeMessurier had regularly and candidly kept the officers advised of all progress and complications, and on several occasions, the design review team itself met directly with DeFord and Dexter to bring them up-to-date on their findings. The more the bank learned, however, the more nervous they became.

Finally, Dexter, himself trained in construction, informed the engineers that a tenant in the building was in the process of connecting two floors of their office with a staircase. Would it be possible, he asked, to use the staircase puncture as a site for floor strength testing? LeMessurier, Robertson, and Pinky all agreed that it could. Robertson later wrote, "The test did prove to be immensely beneficial—showing a diaphragm shear strength equal to that of the perimeter welding."[11] In other words, no remediation of the floor trenches was required.

LeMessurier interpreted the conclusion of the trench header issue as full vindication of his Citicorp Tower structural design. "The story of this problem is important to demonstrate that the building as actually built stands up to extreme and careful scrutiny with the principal exception of the 45° wind problem," he later wrote.[12]

Robertson was a bit more reticent in his appraisal of the issue. "Words of caution should be voiced at this point," he informed the bank:

1) the diaphragm, shear loads, as calculated, are almost surely a measure of the lower bound of loads that will exist in the structure—and larger loads may well exist; and
2) the entire stability of the building is dependent on the shear strength of the steel deck through this trench header.[13]

Years later, when asked about Robertson's design review of the Citicorp Tower, LeMessurier responded, "Look man, nothing is perfect out there. . . . Any time a building is closely examined for any reason whatsoever you uncover a hornet's nest." "Sometimes we spend an awful lot of money for belt and suspenders which are not necessary."[14]

◢

Work on the Citicorp Tower continued through much of the summer. In accordance with LeMessurier's floor-by-floor prioritization plan, by August 18 the reinforcing welds on the twenty-eighth and thirtieth floors were complete and about 60 percent finished on the twentieth and twenty-second floors. Work on the fourteenth floor was well under way and just beginning on the thirty-sixth and thirty-eighth floors. By the end of the month, the columns of the central core had been reinforced pursuant to Yura's specifications, and the trench header tests were complete. Welding on the wind-bracing system was progressing without major difficulty or controversy.

And then the weather changed.

18

ELLA

"Most tropical disturbances dissipated over the open sea. They collided with powerful winds from the west that dipped from the middle latitudes and blew the tops off their thunderheads. They encountered pools of cold water. They entrained so much dry air they lost their passion. Their pillars of smoke and light became mist. Most of the time. Occasionally they became killers."[1] As Erik Larson observed in his debut book, *Isaac's Storm*, chronicling the devastating hurricane that demolished much of Galveston, Texas, in September 1900, the factors that determined potency or dissipation of weather systems remained elusive even into modern times. While the oft-cited theory that a butterfly fluttering its wings in one part of the world could eventually cause a hurricane in another may be more poetic than literal, the so-called butterfly effect, embraced by many scientists and meteorologists, suggests that even small atmospheric changes may in fact have significant impacts on weather.

In the early 1950s, Edward Norton Lorenz popularized the branch of mathematics and science called "chaos theory" dealing with the sensitivity of complex systems to "initial conditions" or system starting points. Lorenz theorized that even slight variations in temperature, humidity, and surface conditions could produce vastly diverse outcomes in weather patterns over time.

"Deterministic chaos," as the concept was often called, challenged the traditional Newtonian belief that small changes in conditions would result in correspondingly small effects. Lorenz's work contributed significantly to the understanding of the unpredictable nature of nonlinear systems—and the consequent realization of the near impossibility of accurate long-range weather forecasting.[2]

Whether the Atlantic hurricane season of 1978 was affected by the flutter of a butterfly's wings or the vagaries of atmospheric anomalies, by late August "the unpredictable" that LeMessurier and others so dreaded awakened over the churning waters of the North Atlantic.

◢

"My mother always said to me when you're under stress the problems show up in your teeth."[3] By the end of August, the pressure of the Citicorp Tower engineering crisis was finally taking a physical toll on William LeMessurier. As his mother warned, he began feeling discomfort in the lower right side of his jaw. Deeply involved with his work and hoping the pain would resolve, he did his best to ignore the problem and carry on. But the pain only got worse with time.

As LeMessurier toiled through the last days of the month, Robertson's weather forecasters detected a weakening and nearly stationary cold front decaying in the central North Atlantic. Fed by warm tropical waters and favorable atmospheric conditions, the seemingly benign system rapidly morphed into a "cyclonic turning of low cloud elements." On August 28, satellite images showed the storm gaining strength about 480 nautical miles southeast of Bermuda, and thirty-six hours later, meteorological data confirmed tropical depression status—and a west-northwest heading.[4]

As word of the storm reached the engineers, the team escalated its disaster readiness posture. Labor Day weekend was approaching, and the city was bustling. By 6:30 on the morning of August 30, the engineers were already gathered at the Waldorf command center and evaluating possible evacuation scenarios. The welding work was progressing, but, as LeMessurier observed, "we weren't quite ready."[5] Though there was no actual panic among the team, they remained in constant communication with the forecasters and tracked the storm with mounting distress.

Throughout the day on the thirtieth, the storm intensified, and by early evening, it had been upgraded to Tropical Storm Ella—the fifth named storm of the 1978 Atlantic hurricane season. The 9 p.m. National Weather Service bulletin warned, "SATELLITE PICTURES INDICATE ELLA IS CONTINUING TO STRENGTHEN AND COULD BE A HURRICANE WITHIN 24 HOURS."[6]

Flight reconnaissance missions and ship reports confirmed the weather service alerts, and, as warned, Ella was upgraded to hurricane status late on August 31 and was moving steadily toward the North Carolina coast at ten to fifteen miles per hour.

LeMessurier was not waiting for Ella to arc north toward New York. That possibility clearly existed, and he feared that the Citicorp Tower may finally be put to the ultimate test. With a deeply impacted molar throbbing in his mouth, he attended a series of hurriedly scheduled meetings between the engineering team and the bank to discuss the current health and status of the building. The question on everyone's mind was whether the tower could survive a direct hit from Ella.

When LeMessurier had created his priority repair analysis, he had absolutely assumed that disaster could strike at any time; hence, the portions of the building most vulnerable to intense wind loads had already been addressed. Now, as fearsome weather approached, the bank officers once again nervously asked if any changes should be made to the list of priorities. Confident in his calculations, LeMessurier told them, "No change in plans. You've got to get these joints welded in time."[7]

LeMessurier surmised that if the TMD could be depended on—and he was optimistic that it could—the building, at that point in time, could most likely survive a two-hundred-year storm. The event that most concerned him, and the one that could, in his view, still destroy the tower, was a storm that entered the Narrows of the Hudson River with its eye just west of Manhattan, advancing northward and unleashing the full force of its southeast winds on the Citicorp Building. He feared that Ella could be that storm.

Furious efforts to develop and complete the trench header test were still ongoing as the hurricane rumbled along the Atlantic seaboard. Fearing the consequences of delay, the engineers insisted that "Pinky" Pinkham work well into the night of the thirty-first. He finally began testing of the Citicorp flooring system at 11 p.m. The favorable results brought a metaphorical port in the raging storm.

By Friday afternoon, September 1, LeMessurier had canceled his plans to spend the holiday weekend in Maine with his family. As work on the tower painstakingly advanced, Ella continued to strengthen. With barometric pressure of 959 millibars and winds exceeding 125 miles per hour, the storm barreled northwestward, and at 4:50 p.m., the National Weather Service issued a "Special Request" to the broadcasters of eastern North Carolina and southeastern Virginia to remain on the air beyond authorized hours in order to provide information to the public on the approaching danger.[8]

Robertson's private weather forecasters methodically tracked the storm's progress and provided the team with near-hourly reporting. As Ella advanced toward the Outer Banks, its central pressure gradually began to rise, and the forecasters reported a slight weakening. For the next twenty-four hours, the storm essentially stalled and drifted slightly northward. With the hurricane's strength and ultimate bearing uncertain, the engineers and bank officers held their collective breath. "Everybody was sweating blood," LeMessurier recalled.[9]

By Saturday, September 2, a ridge of high pressure had enveloped the Northeast like a protective cloak. The weather services observed a slight shift in the storm's direction to the north and east and, soon after, a swift acceleration away from the Atlantic seaboard. Though Ella would greatly intensify through the coming days and ultimately slam the Maritimes with peak winds of

140 miles per hour—the fiercest storm ever to assail Canadian waters—it would spare the United States, the Citicorp Tower, and William LeMessurier untold anguish.

◢

"Saturday was the most beautiful day that the world's ever seen," LeMessurier recalled, "the skies sunny and crystal clear."[10] The holiday weekend had accomplished what Ella had threatened: a near evacuation of the city.

LeMessurier had the day to himself in an empty metropolis, and for the first time in months, his mind wandered from ethical principles and structural dangers. Awash with relief, he took a taxi to the Cloisters, an art museum in Upper Manhattan endowed by John D. Rockefeller and styled after a medieval European monastery. He strolled silently among the calming gardens and ancient architecture overlooking the Hudson River.

He began to breathe again.

19

"EVENTS NOBODY ENVISIONED"

Following the death of Roxcy Bolton in 2017, she was described as "a pioneering and tempestuous Florida feminist." Credited with founding the nation's first rape treatment center, the "scrappy, card-carrying member of the Daughters of the Confederacy" was also the central force in dismantling the long-held practice in the meteorological establishment of assigning only female names to hurricanes and tropical depressions. "Women deeply resent being arbitrarily associated with disaster," she said in 1968. Later, she glibly insisted that "him-icanes" be added to the American lexicon.[1] Due in part to Bolton's relentless efforts, the gender-offensive practice, steeped in maritime tradition, ended with the 1978 hurricane season. Ella was one of the last of a vanishing class.

Despite the fury in the North Atlantic during the first days of September 1978, the weather in New York remained fairly calm throughout much of the summer, and the corrective welding on the tower continued unabated. As the design review progressed, it revealed further minor deficiencies in the tower, some of which required additional structural repairs beyond the strengthening of the chevron bracing members.[2]

By September 11, most of the high-priority welds had been completed, and LeMessurier estimated that the building could now withstand a seven-hundred-year storm without help from the TMD. On that day, with much of the danger passed, he convened a meeting at the bank with Robertson, DeFord, and Dexter to discuss termination of the weather watch, TMD monitoring, and the emergency evacuation plan. Robertson requested some additional time to consider the matter but was generally in favor of the move. On September 13, after reviewing the progress of the ongoing design audit and repair work, he notified the bank of his agreement to end all emergency measures.

On the same day, the bank's general counsel, Hans Angermueller, forwarded a certified letter to the engineers and architects on the project notifying them that Citicorp intended to seek compensation for all damages suffered as a result of the engineering flaws in the building. "We expect reimbursement from you," he wrote, "for all of the costs and expenses, direct and indirect, which we incur in connection with or attributable to the defects, remedying them and ensuring the safety of the office building. We will inform you of the total amount thereof as soon as the work has been completed."[3]

"The insurance company knew perfectly well," recalled LeMessurier, "that sooner or later somebody's gonna come around and want some money."[4]

⬤

Even as the engineers and architects were busy solving the tower's engineering flaws, they each recognized their potential legal peril. As early as mid-August, lawyers were already formulating defenses and framing strategies.

Though Max Edelman's law firm represented LeMessurier's insurance carrier, it was evident that the engineer would personally require separate counsel. As talk of negligence and potential lawsuits percolated, Tim Francis, then president of Sippican, decided to "lawyer up." Believing that the mess would ultimately find its way into a courtroom, he hired a litigation attorney, Dick Lovell, from the one-hundred-year-old Boston law firm Rackmann, Sawyer & Brewster to represent both the corporation and LeMessurier himself.

In a brainstorming session held at Sippican's Marion, Massachusetts, headquarters on August 21, joined by Edelman on conference call, the elements of a legal defense began to emerge. Al Romaneski, who attended the meeting, later noted, "Max gave his general impressions of the case, stating that Bill LeMessurier had acted in a highly responsible way, and the favorable reaction of the owner was largely due to Bill's efforts."[5] Edelman informed the group that he had also requested a New York lawyer, Peter Morrison, to assist with local investigations, research, and law.

At the end of the meeting, Romaneski quietly handed Dick Lovell a notebook containing all of the documents and correspondence relating to the engineering crisis. He explained that because of the extremely sensitive nature of the matter, there were only two copies of these documents in existence. Edelman likewise stressed that absolute discretion was critical. There was one legal file concerning the case in his office, and that file, he claimed, was always kept under lock and key—and he used only one trusted typist for documents pertaining to the matter. "Max emphasized that confidentiality is very important in this case, and urged all concerned to maintain confidentiality in a very strict sense and degree."[6]

The following morning, LeMessurier met with Peter Morrison over breakfast at the Waldorf to discuss potential liabilities and theories of the case. Morrison was particularly interested in the news clippings about the Citicorp Center and the legal relationship between LeMessurier and the Office of James Ruderman. Under the law, LeMessurier's professional missteps could potentially be imputed to others.

On Morrison's advice, LeMessurier immediately met with his joint venture partners, Murray Shapiro and Leo Plofker, and provided them with details of the corrective work already under way at the tower. He apologized for failing to keep them advised of the situation and for not involving them in the solutions that had been generated. Claiming that the urgency of the situation had prevented him from doing so, LeMessurier promised to consult and advise them as the work continued.

The overture fell on deaf ears. By the end of the year, Plofker and Shapiro would also retain counsel and personally inform Romaneski that the Office of James Ruderman disclaimed any and all liability for the Citicorp Tower structural repairs. The error, Shapiro contended, was entirely LeMessurier's since the Office of James Ruderman had no input in the original design of the wind-bracing system or its repair. He added that in failing to even consult the firm about the work to be undertaken, LeMessurier had, in effect, admitted that the mistake was his alone.

◢

Citicorp's September 13 claim letter caused an instant stir among the engineers and their lawyers. LeMessurier was in Cambridge when it arrived in his New York office, and Stanley Goldstein called and read it to him over the phone. Concerned but not particularly surprised, LeMessurier asked Romaneski to notify the attorneys.

On September 25, Max Edelman convened a meeting at his New York office with the various LeMessurier, Sippican, and Ruderman lawyers and their clients to discuss how to respond to Citicorp's formal demand. LeMessurier himself was still dealing with severe tooth pain, and Romaneski appeared at the meeting on his behalf. Though the bank had not yet officially assigned a specific dollar amount to its claim, DeFord and Dexter had indicated in casual discussions with LeMessurier estimated costs between $4.1 million and $4.5 million. Though LeMessurier had not aggressively disputed the assertion at the time, he personally believed the estimate to be grossly overstated and later conveyed to his attorneys his strong belief that the matter could be settled within the limits of available insurance coverage.

At the meeting, it was revealed that Sippican and LeMessurier maintained professional liability insurance coverage of $2 million, with a $100,000 deductible, and Ruderman's office carried $1 million of coverage, with a $15,000 deductible. Beyond those limits, any stipulated or court-ordered awards would have to come out of their own pockets.

Ruderman's attorney, Milton Gould, whose client list during a long and storied career would include prominent figures such as Aristotle Onassis, Leona Helmsley, and Ariel Sharon, remained stoic throughout much of the meeting. Though he knew his client may well incur some legal liability by virtue of its mere membership in the joint venture, Gould had still taken the position that LeMessurier had caused the problem, not Ruderman. But for now, he believed cooperation served his client better than confrontation and decided to approach the problem collectively. Toward the end of the meeting, the well-connected Gould told the gathering that he was scheduled to have dinner with Walter Wriston in the next few weeks and that he might be able to determine more about the bank's attitude in the matter at that time.

After learning about the additional work to the tower prompted by the ongoing design review, Edelman pointed out that the overall corrective measures were much more extensive than he had previously understood. For defense purposes, he requested that LeMessurier prepare a detailed report documenting the work performed and indicating to what extent it had resulted in the "betterment of the building." The report, Edelman advised, was to be marked "Privileged—For Counsel Only."[7]

In the meantime, it was agreed that each lawyer would reply to Citicorp's letter with a simple acknowledgment of receipt by their respective clients.

◢

The pain in LeMessurier's tooth was unrelenting. On September 27, he underwent oral surgery at Mount Auburn Hospital in Cambridge for the removal of two molars, one impacted and the other cracked. He was barely able to talk for days.

◢

By the end of October, most of the corrective work on the tower had been completed, and LeMessurier was satisfied that the building was as strong and safe as any other high-rise structure in the world. Les Robertson later wrote, "Citicorp Center has been subjected to structural repairs of unprecedented variety and magnitude."[8]

The design review engineers returned to Cambridge on September 23 to continue their work closer to their families, though they attended weekly review meetings at Robertson's New York office. On November 21, the completed document was signed by the three engineers and delivered to Robertson. The extent and breadth of the final audit, which would yield over twenty-four hundred pages of calculations in four volumes and actually expand the areas of structural concern in the tower, quickly dispelled any notion of bias or unprofessionalism in the review team.

⊿

When LeMessurier first created a confidential office file for the Citicorp Tower structural problem, he assigned it the name "Project Pandora," presumably alluding to the mythological Pandora's box that the matter represented. On August 7, he changed the name to "Project SERENE"—an acronym for "Special Engineering Review of Events Nobody Envisioned," on the advice of an associate. The word "Serene," they agreed, projected a more subtle and calming impression.

Though every memorandum, report, document, and letter pertaining to the crisis would fall under the mantle of "Project SERENE," the "privileged" statement requested by Max Edelman to itemize and evaluate the corrective work completed on the tower would become, in its final form, synonymous with the title.

LeMessurier continued to experience severe pain after his oral surgery as well as a deep bone infection that would ultimately land him in the hospital through the Thanksgiving holiday for intravenous antibiotic treatment. With Al Romaneski's assistance, he continued to work through the pain. They prepared various iterations of the statement that culminated with a sweeping thirty-page narrative, aptly titled "Project SERENE." It was a cathartic exclamation point on a trying and emotional chapter of LeMessurier's career.

Originally intended as a mere itemization of completed work, LeMessurier's final document would morph instead into a comprehensive chronicle of the Citicorp Center and the structural crisis that enveloped it. He included a detailed account of the design and construction of the building, including the formation of the LeMessurier-Ruderman joint venture, the production of construction documents, and design of the wind-bracing system and TMD. "The final arrangement of framing members and their proportioning for rigidity in high winds was accomplished in an economical and even elegant fashion," he wrote. "It later developed that inherent in the unusual framing system was a sensitivity to wind forces acting at a 45° angle to any of the building's four faces."[9]

LeMessurier described the change from full-penetration welds to bolts for the tower's bracing connections and the manner in which he discovered the vulnerability of those connections to quartering winds. "Review . . . showed that only perpendicular winds were considered," he admitted.[10] "Based on the worst case . . . the building was in danger of collapse in a storm which might occur as often as once every sixteen years."[11] He then chronicled his disclosure of this vulnerability to the building designers and the bank officers and outlined his strategy for repair of the problem.

The critical input of Alan Davenport and his wind-tunnel staff and Robertson's far-reaching list of expert technicians who assessed the building and its structural flaws suffused the report, and LeMessurier acknowledged—and sometimes critiqued—Robertson's cautious approach to the building's assessment and repair. "Any review of any building by a second professional is bound to bring out differences in philosophy," he wrote.[12]

LeMessurier recounted the extensive meetings between the engineers, architects, experts, insurers, lawyers, and bank officers, and he examined the decisions made and the emergency measures formulated. His candid description and personal assessment of the design review conducted by his own engineers under Robertson's supervision dominated much of the narrative, though he added, "Several . . . concerns raised by Robertson (in good faith) have proven to be unimportant after detailed study."[13]

"It may be positively stated," wrote LeMessurier near the end of "Project SERENE," "that the building is in a very sound and reliable state." He claimed that the corrective work, once complete, would ensure that the structure could withstand a one-thousand-year windstorm.[14]

◢

At its core, "Project SERENE" was intended as a statement of legal defense—and solely for the eyes of his attorneys. In preparing the document, LeMessurier remained convinced that any litigation brought by Citicorp was defensible—though the last thing he wanted to do was go to court. He insisted, for legal purposes, that sound engineering, professional guidelines, and applicable code required consideration of only perpendicular winds in the design of the Citicorp Tower. "Design for 45° winds is not a standard of practice followed by the majority of tall building designers. It is certain that very few offices routinely consider the problem," he wrote. "Numerous witnesses could be found who would support the contention that normal good practice is based on perpendicular winds."[15]

But then, rather astonishingly, LeMessurier confessed that he had been aware of the significance of quartering winds *for years* and acknowledged that a defense based on a regular engineering practice with potentially catastrophic outcomes is not very compelling. "Nature does not know the intent of building designers. Winds are equally probable from the Northeast as they are from the North or the East."[16] Instead, he argued, the focus of their defense should be on the net benefits derived by Citicorp as a result of the pioneering and cost-saving original design of the tower.

He pointed out that the initial studies contemplated by the design team envisaged a "table" of sorts on top of the bottom columns that would extend over St. Peter's Church. That solution would have resulted in a conventionally framed building placed on top of the table and would have required an additional $5–6 million in steel costs. The bank, he wrote, was fully prepared to carry that additional expense in the absence of other alternatives. Writing in the third person, he continued, "LeMessurier might have stopped with this conventional and expensive solution. Fortunately, he did not. Instead he developed a design specifically tailored to the unusual site conditions. Rather than building a six million-dollar table he organized the wind and gravity load carrying systems into the now famous 'Chevron' system of eight story diagonals which turn the whole building into a very efficient 48-story table. The reduction in steel quantity from this innovation was breathtaking."[17] He estimated an additional cost savings to the bank of $10 million in reduced construction costs due to his novel bracing system.

Although Citicorp had not yet, at the time, assigned a dollar figure to its formal demand, LeMessurier believed it would be somewhere around $4 million. He maintained that those costs, large as they appeared, were far less than the "very real savings achieved" by his novel structural design. "It is almost the nature of new systems to have unforeseen problems which can only be disclosed by experience and evolution. The problems at Citicorp Center, however real, were only potential. Not the slightest physical evidence of problems had shown up prior to LeMessurier's disclosure. The owner was not damaged in any way by bad performance of the structure. . . . The net result of the entire experience, painful as it was," he averred, "is that Citicorp still has an economical building which is now known to be safe and reliable and which is the outstanding architectural and commercial success in New York."[18]

Further dismissing the bank's damages, LeMessurier justifiably congratulated himself on his admirable response to the crisis:

When LeMessurier told the Bank of the very real potential problems he also was ready with solutions. Rather than leaving the owner in a helpless state of shock, the design team responded at once with effective planning and direction of the corrective procedures. Together with the owner's own brilliantly organized staff and the cooperation of the City and Contractors, the team operated with unbelievable speed to make the building safe. Meanwhile every possible precaution was taken to protect the interests of the public from possible disaster. Only six weeks after the disclosure to the Bank the emergency was officially over. Thanks to excellently managed public relations, . . . neither the owner nor the building has suffered any serious loss of reputation. The six week period was filled with risk and danger but goodwill and luck carried the day.[19]

LeMessurier concluded "Project SERENE" with a rhetorical flourish. "The final question is simple," he wrote. "Viewing the entire story of the designers' professional behavior in serving the interests of Citicorp, did Citicorp suffer net damage? Or did they, after six weeks of agony and peril, end up with one of the best real estate investments in the world?"[20]

⊿

As LeMessurier shaped and hewed the narrative of "Project SERENE," Robertson's design review team was wrapping up its investigation of the tower. Prior to completion of the audit, LeMessurier had been optimistic about its pending conclusions. Most of the corrective work advocated by the team had been completed, and the building was clearly out of danger. "It is expected that by November 20 all investigation will be completed and that the existing structure will be found entirely safe and reliable," he wrote.[21]

He was only partially right.

⊿

For the next several months, the architects, engineers, insurers, and lawyers held their collective breath while Citicorp calculated the amount of its final claim. Finally, on May 3, 1979, Hans Angermueller issued another demand letter enumerating costs of $3,549,308 in addressing the tower's structural defects. He cautioned that the matter was ongoing and that further expenses were expected.

And, in an apparently deliberate move to ratchet up the legal tension, Angermueller referred to the building's designers in his letter as "the defendants."[22]

⌁

In early July 1979, Robertson delivered to the bank his own "Structural Investigation of Citicorp Center," providing the historical backdrop of the crisis and explaining in detail the curative measures that had been recommended and undertaken at the tower. The report also provided a thorough description of the design review team's work and itemized the repairs made to the building as a result of its audit.

At the outset, Robertson acknowledged that the tower had indeed been in danger. Referring to the weakness of the bolted bracing members, he wrote, "It is the opinion of [Skilling, Helle, Christiansen, Robertson], and we believe it to be the opinion of LeMessurier, that failure of one of these connections would likely result in the progressive and total collapse of the building."[23]

Though Robertson and LeMessurier clearly had their differences, their professional relationship throughout the crisis remained largely cordial and cooperative. In Robertson's report to Citicorp, he was careful to depict LeMessurier as an honest broker in the ordeal. "We believe," he told the bank, "that the posture of Mr. LeMessurier is properly demonstrated by his having brought his concerns to Citibank and in his insistence that recognized experts be brought in to deal with the problem."[24] Later, Robertson told a journalist, "Bill was very forthcoming. . . . Maybe all of us would not behave like that. . . . I have a lot of admiration for Bill because he was very forthcoming."[25] Privately, however, Robertson's approbation did not extend to LeMessurier's judgment and style.

On August 8, in the midst of the crisis, LeMessurier had informed a gathering of city officials that the tower had been constructed in strict conformity with the New York City Building Code.

Robertson, who was also at the meeting, held his tongue. He now felt duty bound to inform the bank that LeMessurier's statement was not entirely true. "Now, on completion of the Design Review, we can state with confidence that the structural system for the Citicorp Center did not conform to the requirements of the Code," he wrote in his report. "Even more important," he continued, "now that the repairs are complete, the structural system *will* meet with standards of sound engineering practice but *will not* always meet the specific requirements of the Building Code."[26] Robertson acknowledged that practicality simply did not allow strict code compliance in every instance of repair, but he left the final decision of whether to accept that posture of forbearance with the bank's management team. As to the reliability of LeMessurier's statement to the city officials, Robertson wrote, "As a part of this report we feel the need to counsel Citibank to the end that the [Building] Department may need to be informed of these areas wherein the structure does not conform to the requirements of Code."[27]

Though Robertson's report clearly confirmed the bank's understanding that "strict compliance with the requirements of the Building Code was not required so long as sound engineering practice was followed in every respect," just days after receiving it, Robert Dexter forwarded a formal letter to LeMessurier alleging that the Citicorp Center, as originally designed, failed to conform in every respect to the New York City Building Code.[28] The letter further claimed that "while the Center is now as safe as comparable buildings and that, while the structural modifications made . . . do follow sound engineering practice, the Center does not in every respect, conform to . . . Code."[29] It was, perhaps, the bank's way of placing the issue "on the record" in preparation for litigation—or even a cynical effort to gain leverage in the negotiations that it knew would soon follow.

◢

In mid-November 1979, Citicorp made its final demand for payment. The amount, $4,300,391, included every conceivable cost from construction and lost rent to attorney's fees, consultants, officers' expenses, and even food services.

Hoping to avoid the acrimony of litigation, LeMessurier approached Northbrook Insurance Company shortly thereafter and asked permission to personally attempt settlement of the matter—without lawyers to muddle the waters. Max Edelman no doubt bristled at the suggestion, but after studying "Project SERENE," he was persuaded that LeMessurier may, in fact, have been better equipped to handle the matter than most anyone else. Northbrook, which also wrote the liability policy for the Office of James Ruderman, granted the request. Of the $3 million of total coverage, LeMessurier was given authority—subject to final approval—to settle the claim for $2 million.

Once the bank's final dollar demand had been digested, LeMessurier notified DeFord and Dexter that he would like to personally meet to discuss settlement of the claim. The men had successfully navigated the stress of original disclosure and repair of the building; now, LeMessurier implored, they should sit down in the same spirit of cooperation and resolve their differences. He explained that he would not be bringing an attorney, and he asked that they do the same.

During the discussion, LeMessurier made many of the points he had raised in "Project SERENE." He assured the officers that the novel design of the building had resulted in a net gain despite the now-resolved structural flaw. No court, he promised, would penalize him for failure to consider quartering winds in the tower's design. It was simply not within the requirements of sound engineering practice. He told them that he had, nonetheless, been given authority by his insurance company to settle all claims for $2 million.

DeFord and Dexter balked. They were unimpressed—perhaps even offended—by the suggestion that a structural design that ultimately cost them over $4 million to correct somehow placed them in a more *favorable* position. They reminded LeMessurier that the building still had not been brought into strict code compliance and, forebodingly, asked whether the city officials had yet been notified of the fact. They pointed to the itemization of damages that was included with Angermueller's latest demand and insisted that the bank was entitled to recoup every dollar expended in dealing with the crisis. After all, they told LeMessurier, Citicorp had not caused this problem—he had.

Despite the men's bluster and angst, LeMessurier still sensed a subtle willingness to resolve the matter. He suspected that Walter Wriston and the bank's upper management were probably not interested in bringing the matter to court. Citicorp had paid an estimated $190 million for its signature building in Manhattan, and the expense of the structural defect was, considering the bank's overall financial footing, a trifling matter. He was confident that the officers would want to avoid the probing press coverage and embarrassing public relations that legal proceedings would bring. The meeting with DeFord and Dexter failed to resolve the case, but LeMessurier was confident that it opened the door for further discussions.

On September 29, 1980, LeMessurier again met with Citicorp officials to discuss settlement. Nearly a year had passed since his first attempts at resolution with Dexter and DeFord, and attitudes had clearly softened. He had remained in contact with the men, and by July 1980, he was able to report to Carl Sapers, "it is clear that the goal of both of us is to settle without litigation."[30]

Though LeMessurier still handled the negotiations, this time he was accompanied by Peter Morrison, one of Northbrook's attorneys. Progress was made during the discussion, and finally the general parameters of an understanding began to emerge. Over the next several weeks, the lawyers amicably brokered the finished details of the settlement: in return for a full release of all claims against the entire design team, the insurance company would pay the bank $2,050,000—$1,050,000 from Sippican's policy and $1 million from Ruderman's policy.

By the end of the year, the Citicorp engineering crisis was finally over.

"Our behavior," LeMessurier later said, "prevented one of the greatest insurance disasters of all time."[31]

⊿

"What does 'professional' mean?" LeMessurier once asked. "Doctors, lawyers, engineers, architects . . . have a social obligation to society, in return for the privileged status of getting a license, being addressed with distinction. You are supposed to be self-sacrificing and look beyond the interest of your client to society as a whole. This is an obligation on you. It means if you made a mistake you must serve your client's interest and the public's interest above your own interest. There's no choice really."[32]

Of the Citicorp engineering crisis, the journalist Joe Morgenstern wrote years later, "it produced heroes, but no villains."[33] Even Leslie Robertson would reflect, "All parties behaved in an absolutely exemplary manner. . . . All of the work that was done was first-rate and there were no holds barred in terms of making the project the best it could be. . . . I thought [Bill] was very composed throughout the entire event. . . . I think he was very cool and in charge of himself and the situation."[34]

The potential dangers to the Citicorp Tower and to the city at large were incalculable. As the world would sadly learn many years later, a collapse of a major skyscraper without alert in the heart of Manhattan could cause unspeakable tragedy and untold loss of life. "There were no evident signs of distress, nor could any be expected until the last horrible moment," LeMessurier later reflected of the Citicorp Tower. "Those brittle failures, when bolts snap, don't give any warning. They're there one minute and gone the next. So this is a crucial case, a special one."[35]

At risk to his reputation, his career, and his livelihood, he confronted the consequences of inaction, and he confessed the insidious, lurking danger of structural failure. He met his responsibility as an engineer—as a "professional."

And yet, through the years and for the rest of his life, William LeMessurier would reflect on the enduring lessons of the crisis.

Could he have prevented the mistakes of the past? Had he truly served his colleagues, his clients, and his community with honor and principle?

And he wondered in his quiet moments, as most structural engineers do, what other buildings are silently biding their time, waiting for *their* one-thousand-year storm, *their* one-hundred-year storm.

Their sixteen-year storm.

EPILOGUE

By the time the New York City pressmen's strike had been settled, most of the structural repairs to the Citicorp Tower were complete. The city newspapers were back to covering elections, the Yankees, and rising gas prices. With the exception of Rippeteau, the hard-nosed reporter from *Engineering News-Record* whom LeMessurier did his best to elude following the crisis and who herself was in the process of being hired away by *Business Week*, the media had ignored the curious illumination of the tower during the summer nights of 1978.

Fearing negative personal or business repercussions, most of the people involved with the matter chose to remain silent, and the story receded from the public consciousness—to the extent that it ever even occupied it. Gradually, LeMessurier began to discuss the Citicorp story in small circles, first with his students to explain the unique engineering and ethical challenges encountered and then in off-the-record, closed-session academic conferences to one or two professional audiences. Though whenever he discussed the topic, people appeared genuinely interested and eager to learn more, for the most part—and to the relief of many—awareness of the crisis gradually waned.

In 1993, a freelance journalist named Joe Morgenstern stumbled on a *New York Times* article about the structural frailties of New York's East River bridges and, always on the lookout for a block-buster story, began speculating about the possibility of a fictional made-for-TV disaster film questioning what would happen if a fatal flaw were detected in the Brooklyn Bridge.[1] Morgenstern, who would later be awarded a Pulitzer Prize for criticism while writing for the *Wall Street Journal*, met some old friends and their son-in-law, Barry Price, for dinner one evening in Woodstock, New York. Learning that Price had recently graduated from the Harvard Graduate School of Design, Morgenstern told him about the TV concept he had been considering and asked whether the Brooklyn Bridge *could* actually fail. Responding that perhaps under the right circumstances it could, the young architect recounted a compelling story involving the Citicorp Tower in the late 1970s that he had studied in a structural engineering class at Harvard. The professor, of course, was William LeMessurier.

Morgenstern found himself captivated by the story but wondered if it was actually true. In 1982, he had written a piece in *New York* magazine about the iconic architect Frank Gehry, and the two became good friends. He reached out to Gehry and recounted what Price had told him over dinner about the Citicorp Tower. "Sounds like bullshit," Gehry said. He found it hard to believe that such a near disaster could be hidden for all those years but said he knew someone at the company that constructed the tower and would check. According to Morgenstern, Gehry called back ten minutes later. "I'll be damned. It's true," he said.

Morgenstern thought the obvious home for a "technical as well as dramatic" piece about the Citicorp engineering crisis would be the *New Yorker*. He contacted the magazine's lead editor, who immediately expressed interest and authorized an assignment. Then, "with trepidation," he called William LeMessurier at his Cambridge office. As Morgenstern recalled, "After asking what I wanted, his secretary put me through. The first words out of his mouth were, 'I've been waiting for this call since 1978.'" "He was not reticent, to say the least," said Morgenstern. "His eagerness to tell the story to his students, and to have it told to a national readership, was prompted by his desire to be seen as a Man Who Did the Right Thing—a fair self-assessment in the end." Morgenstern's ultimate piece, "The Fifty-Nine-Story Crisis," appeared in the May 29, 1995, issue of the *New Yorker*, and beyond the fusillade of praise from the general public, it soon became required reading in architectural and engineering ethics classes across the country.

As LeMessurier had perhaps hoped, his actions in disclosing and resolving the crisis as depicted in the piece were met with almost universal applause. As one reader commented, "It makes one realize that the term 'professional ethics' is not always the oxymoron the cynics among us think it is."[2] Shortly after publication of Morgenstern's piece, the book acquisitions editor of ASCE Press—the publishing arm of the American Society of Civil Engineers—wrote to LeMessurier expressing interest in developing his account of the crisis into a full-length book. "We think that this story, . . . written from the engineer's point-of-view, would be very appealing to our readers," wrote the editor.[3] And several months later, Richard Korman, writing for *Engineering News-Record*, reported, "It may be structural engineer William J. LeMessurier's destiny to be known for the error he confessed rather than for the skyscrapers he designed. At no time in his 69 years has LeMessurier ever been better respected."[4]

But Morgenstern's piece also stirred controversy. Stanley Goldstein, who assumed control of LeMessurier's New York office only after the structural design of the tower was complete and construction was well under way, objected to any insinuation that he had somehow presided over the project. "'I was not the project manager when the change [to bolts]' was made," he told a gathering of the New York Society of Professional Engineers on October 12, 1995.[5] And he complained to *Engineering News-Record* that as a result of the *New Yorker* article, he had endured "comments from clients and others associating him with the problems."[6] Though he could not say for certain that LeMessurier was personally aware of the change to bolted connections, Goldstein insisted that the Cambridge office had been informed of the change. Bob McNamara, LeMessurier's project engineer in the early design phase of the tower, claimed, according to *Engineering News-Record*, that, contrary to the assertions in Morgenstern's piece, LeMessurier knew of the design change from welds to bolts.

LeMessurier, speaking for himself, claimed that he tried to protect his New York team while collaborating on the *New Yorker* article. He maintained, however, that he had not been informed of the change to bolts and did not know who generated the calculations for the connections as modified. He confessed only to not providing sufficient oversight to his personnel during the project.[7]

While LeMessurier admirably disclosed to Citicorp the engineering blunders his office had committed with the tower, at great risk to his career and reputation, he was not without his detractors. Once the full story of the averted catastrophe was publicized in the *New Yorker*, some people insisted that his actions revealed specific ethical lapses and violations of professional responsibility.[8] In an attempt to avoid panic in the community and professional embarrassment to himself and others, he had been less than truthful in his public statements. "We had to cook up a line of bull, and white lies at this point are entirely moral," he had said.[9]

A robust conversation has ensued in the design industry and the classroom questioning whether LeMessurier had violated professional codes of ethics in failing to accurately disclose to the public the full extent of the threat presented by the Citicorp Tower engineering flaw.[10] Of equal concern, according to the critics, was the concerted effort to hide the emergency and evacuation measures that had been developed during the tower's repair. Even after the crisis, LeMessurier had, for all practical purposes, secreted from the public eye the full extent of the near calamity in order to avoid scrutiny and criticism. It was not until sixteen years later, upon release of "The Fifty-Nine-Story Crisis," that the full scope of the looming disaster unfolded. Could the structural design community have benefited from the lessons learned with the Citicorp Tower? Could other disasters have been averted had LeMessurier elected to immediately tell his story? While the goal of avoiding public panic during the crisis was unquestionably reasonable, the tenants, occupants, and members of the community whose personal safety was placed at risk still had the right to know the full scope and imminence of the potential danger in order to make informed decisions. LeMessurier was certainly not alone in these strategic judgments—but he clearly supported them.

Perhaps LeMessurier's greatest offense—and the one that he freely admitted—was the failure to provide adequate direction and oversight to his New York engineers in the development of the Citicorp Center engineering drawings, calculations, and revisions. He would later ask himself whether the crisis could conceivably have been averted had he been more involved in the process beyond the initial conceptual design of the tower. He was never professionally censured for any of these lapses; to the contrary, he was almost universally commended for his disclosure and cooperation throughout.

In the January 1997 issue of the *Journal of Professional Issues in Engineering Education and Practice*, the American Society of Civil Engineers reprinted Morgenstern's *New Yorker* piece in its entirety and opined, "LeMessurier's exemplary behavior—encompassing honesty, courage, adherence to ethics, and social responsibility—during the ordeal remains a testimony to the ideal meaning of the word,—'professional.'"[11]

Justified or not, this remains William LeMessurier's enduring legacy.

◢

Like most people, Diane Hartley was totally oblivious to the engineering crisis that enveloped the Citicorp Center in the summer of 1978. She was not immediately aware of the *New Yorker* article when it came out in 1995, and she had absolutely no inkling that a call from a student in New Jersey had led to LeMessurier's discovery of some structural vulnerability in the tower. In the mid-1990s, she was busy building a career and raising a family. To her, the Citicorp Tower was a distant memory.

That all changed in the spring of 1996. One evening, while Hartley was quietly coaxing her infant son to sleep, her husband suddenly shouted from downstairs. A television program about her "thesis building" was on. Stunned, she quickly turned on the set, and indeed, a BBC documentary titled "All Fall Down" describing the Citicorp engineering crisis was being presented. Watching with astonishment, Hartley gaped when the narrator described how a call from a student in New Jersey had triggered the events to come. "I nearly dropped my son," she recalled.[12]

Hartley had noticed that the program depicted the unnamed caller as "he," and so she wondered whether some other engineering or architectural student had called LeMessurier and induced his actions. After all, she thought, she had never personally spoken with LeMessurier—only with his New York engineers. Later, she dug out her undergraduate thesis, and as she reviewed her work, the questions only deepened in her mind. Unsure of whether she had played any role in what unfolded at the Citicorp Tower, Hartley put the matter out of her mind and went on with her life. The "mystery student" would remain just that.

In May 2003, David Billington, Hartley's thesis adviser and mentor, was honored at Princeton University, and he invited Hartley to the event as his guest. At dinner, he asked if she was aware of the Morgenstern piece about the Citicorp engineering crisis and the claim that a call from a student in New Jersey had prompted LeMessurier's investigation. She told him that she was aware and asked eagerly if he had any idea who the "mystery student" was. Billington explained that there were very few engineering or architectural schools in the New Jersey area and that he was acquainted with the directors and faculty at all of them. He said that he had done some investigation and found no other student who claimed to have made the call—except for her. "It had to have been you," he said.

Morgenstern's article referred to the caller as "an engineering student from New Jersey, . . . [a] young man whose name has been lost in the swirl of subsequent events."[13] Hartley later mused, "Did my inquiries prompt Joel Weinstein himself to realize the errors, and bring the matter to the attention of LeMessurier, who then stepped up to take the hit, . . . with then, perhaps, the 'engineering student' being assumed to be a male?"[14]

Several months later, LeMessurier added to the confusion when he perplexingly told an associate that he never spoke directly with the student and had no recollection of whether it was a man or woman—all contrary to what he had informed Joe Morgenstern in 1993 in his interviews for the *New Yorker* piece.

And then, in 2011, Lee DeCarolis, the architectural student from the New Jersey Institute of Technology, entered the picture. The mystery deepened.

Through the years, Diane Hartley has been heralded as an "indispensable hero," the woman who "saved a skyscraper," and even a "messenger from God."[15] Ironically, at the time of her thesis, she had not even been aware of the primary cause of the problem: the change from welded to bolted bracing connections. But for that change, quartering winds would never have posed any elevated risk to the Citicorp Tower. Even she has remained somewhat perplexed by the onslaught of attention.

In 1988, confronted with what the *New York Times* called "huge losses," Citicorp sold a two-thirds interest in its signature building and a one-third interest in its original headquarters at 399 Park Avenue to Dai-Ichi Mutual Life Insurance Company at a price of $670 million—the latest demonstration of Japanese might in US real estate markets.[16] Dai-Ichi eventually acquired the bank's remaining rights and, in April 2001, sold the tower for $755 million. The majority owner of the building, Boston Properties, announced in 2021 that the loan against the building had been refinanced for the sum of $1 billion.[17]

After the 9/11 attacks, Boston Properties discreetly strengthened one highly vulnerable primary column of the tower located just feet from Fifty-Third Street in an effort to mitigate the risk of structural collapse from a street-level detonation. According to the *New York Times*, the tower, with "a troubled history involving secretive structural repairs," was subjected to "a highly focused, structurally sophisticated and quite expensive effort . . . to buttress . . . against threats that suddenly seem to be everywhere, from the sky to the street, in the post-9/11 world."[18] Two years later, the tower, "one of the premier international symbols of American finance," was designated by the Department of Homeland Security as a high-risk terrorist target.[19]

Following the economic crisis of 2008, in an effort by the owner to attract new tenants and to disassociate itself from the original namesake, the complex that once bore the name of a financial powerhouse became known simply as "601 Lexington."

▲

The events of September 11, 2001, changed the US forever and altered lives in poignant and heartbreaking ways—perhaps none more profoundly than Leslie Robertson's. As the lead structural engineer for New York's Twin Towers, Robertson would personally bear the burden of 9/11. "The World Trade Center was destroyed and so was Robertson," wrote one journalist. "He lost a lot of his joy and spirit. He had to defend himself, because he was attacked, criticized and pressed by other engineers," said the architect A. Eugene Kohn.[20]

Robertson wrote in his 2017 memoir, *The Structure of Design: An Engineer's Extraordinary Life in Architecture*, "My sense of grief and my belief that I could have done better continue to haunt me. Perhaps," he continued, "had the two towers been able to survive the events of 9/11, President Bush would not have been able to project our country into war. Perhaps, the lives of countless of our military men and women would not have been lost. Perhaps countless trillions of dollars would not have been wasted on war. Perhaps, I could have continued my passage into and through old age, comfortably, without a troubled heart."[21]

▲

William LeMessurier died in June 2007 at the age of eighty-one. The author of one tribute, titled "William LeMessurier Builder of Cutting-Edge Structures," wrote, "Bill . . . was known around the world as one of America's most daring tall building designers. [He] was a renaissance man."[22]

Henry Petroski captured the enduring message of the Citicorp Tower crisis and the man who confronted it in his book *To Engineer Is Human*: "Most [buildings] . . . do not fail, of course, but the structural success of another traditional design is no more news than the man who does not rob a bank or does not bite a dog. It is the anomaly that gets the press, and the abnormal that becomes the norm of conversation. Thus, to speak of engineering failures is indirectly to celebrate the overwhelming numbers of successes."[23]

The Citicorp Tower, born of courage and design brilliance, forever bears the mark of an audacious—and all too human—engineer and his willingness to risk all.

ACKNOWLEDGMENTS

It is often said that writing is a solitary endeavor, and for the most part, that is true. But no book project comes together without the assistance, guidance, and patience of others. The backbone of a compelling and truthful narrative is factual data, and the task to reveal that data requires people. It truly does take a village.

In the case of *The Great Miscalculation*, the "village" begins with Joe Morgenstern. His piece "The Fifty-Nine-Story Crisis," which appeared in the *New Yorker* in May 1995, was the first in-depth look at the Citicorp engineering crisis. I am indebted to Joe for sharing with me the details of his research and the transcripts of his in-depth conversations with William LeMessurier, Leslie Robertson, and others. His assistance allowed me to flesh out the narrative with personal accounts that would not otherwise have been available.

I want to thank Diane Hartley for revealing her personal story of discovery to me and for patiently explaining some of the engineering principles behind the Citicorp Tower. Her undergraduate thesis is truly a remarkable and comprehensive academic work, and I thank her for sharing her experience and expertise with me.

Thank you to Lee DeCarolis—the "mystery student"—for sharing his story with me and for shedding some light on his unwitting contribution to LeMessurier's discovery.

For assistance in research, I want to thank Ines Zalduendo, special collections curator of the Frances Loeb Library, Harvard Graduate School of Design, and Kerri Anne Burke, global curator of the Citi Heritage Collection. Also, I am grateful to Richard Henige and Robert McNamara for technical advice and direction and to Holly Lancey for technical drawing assistance.

I owe a special thank you to William LeMessurier's daughters, Claire and Irene LeMessurier, for providing me with some personal insights into their father's character, temperament, and talents. It is not necessarily easy to reveal this kind of information to a stranger for publication, and so I am grateful for their trust and cooperation.

Finally, I thank my agent, Rita Rosenkranz, and editor, Clara Platter, for believing in this story and allowing me the privilege of bringing it to life.

NOTES

PROLOGUE

1 Johnston, "Church Moves in a Procession Up Park," 33.

2 Johnston, "Church Moves in a Procession Up Park," 33.

3 Johnston, "Church Moves in a Procession Up Park," 33.

4 Derr, preface to *Religion and Art*, xx.

5 Lepine, "Upon This Rock," 23.

6 Jackson, Keller, Flood, *Encyclopedia of New York City*, 505.

7 Lepine, "Upon This Rock," 26.

8 Hellman, "How They Assembled," 31.

9 Hellman, "How They Assembled," 31.

10 Campbell, "Citicorp Center," A11.

11 Hellman, "How They Assembled," 31.

12 Hellman, "How They Assembled," 31.

13 Dempsey, Peterson, and Dillenberger, "Ralph Peterson."

14 Harvey, "Jazz Ministry," 158.

15 Dempsey, Peterson, and Dillenberger, "Ralph Peterson."

16 Harvey, "Jazz Ministry," 162, quoting Ralph Peterson, "Art as a Search for Meaning," 1966, St. Peter's Archives.

17 Harvey, "Jazz Ministry," 162, quoting Heschel, *Insecurity of Freedom*, 23.

18 Dempsey, Peterson, and Dillenberger, "Ralph Peterson."

19 Rosen, introduction to *Religion and Art*, 3, quoting Peterson, *St. Peter's*, 1.

20 Dryansky, "Hugh's Masterpiece," 206.

21 Dryansky, "Hugh's Masterpiece," 206. According to the biographer Phillip L. Zweig, Walter Wriston, the chairman of Citibank, personally finalized the deal with the doctor's cooperative during a visit with an eye specialist, Dr. Walter Peretz, who was a member of the group. In order to avoid a massive capital gains tax, Wriston proposed a tax-free exchange of Citibank stock for a controlling interest in the stock of the cooperative. Zweig, *Wriston*, 376.

22 Dempsey, Peterson, and Dillenberger, "Ralph Peterson."

23 Dempsey, Peterson, and Dillenberger, "Ralph Peterson."

24 "John White, Real Estate Agent," 46.

25 Dempsey, Peterson, and Dillenberger, "Ralph Peterson."

26 Dempsey, Peterson, and Dillenberger, "Ralph Peterson."

27 Hellman, "How They Assembled," 35.

28 First National City Corporation was formed in 1967 as a result of the Bank Holding Company Act of 1956 to separate purely banking operations

from the more risky nonbanking activities. This allowed the bank to engage in a broader range of financial services while maintaining a clear separation between the bank's deposits and its nonbanking activities.

29 Citi, "Heritage."

30 Lepine, "Upon This Rock," 36, quoting Ralph Peterson, Comments at the Dedication of Citicorp Center, October 12, 1977, 2–3, St. Peter's Archives.

31 Huxtable, "New Urban Image?," 21.

32 "Office Complex to Hold Church," 33.

33 Peterson, *Life at the Intersection*, 1.

CHAPTER 1. A SKYSCRAPER ON STILTS

1 Ireland, "River Runs through It."

2 Ireland, "River Runs through It."

3 Crosbie, "Hugh Stubbins," N1.2.

4 Boyn and Frohling, *Divided Berlin*, 67. Tragically, the roof of the Congress Hall, now known as the House of the World's Cultures, collapsed in May 1981, killing one person and injuring many. The collapse was caused by rusting of the building's steel core. The building was restored according to the original plans.

5 Crosbie, "Hugh Stubbins," N1.2.

6 Ludman, *Hugh Stubbins*, 85.

7 Landmarks Preservation Commission, "Citicorp Center," 10.

8 Schmertz, "Citicorp Center," 114.

9 Landmarks Preservation Commission, "Citicorp Center," 10, quoting New York City Planning Commission, *New Life for Plazas*, 5.

10 Schmertz, "Citicorp Center," 114.

11 Hugh Stubbins to Henry J. Muller, September 29, 1970, Papers of Hugh Stubbins.

12 Schmertz, "Citicorp Center," 114.

13 Hugh Stubbins to Henry J. Muller, September 29, 1970, Papers of Hugh Stubbins.

14 Mehlman, "Elevating the Urban Environment," 48.

15 Weingardt, "William LeMessurier," 35.

16 Morgenstern, interview with LeMessurier, 1; Ludman, *Hugh Stubbins*, 131n3.

17 Biographical information on William LeMessurier from Henige, "William J. LeMessurier."

18 Ludman, *Hugh Stubbins*, 131n3.

19 Henige, "William J. LeMessurier," 206.

20 Henige, "William J. LeMessurier," 208.

21 Weingardt, "William LeMessurier," 35.

22 See early sketches by Stubbins depicting these scenarios in Schmertz, "Citicorp Center," 116.

23 Morgenstern, of interview with LeMessurier, 7.

CHAPTER 2. "SCARY EXCITEMENT"

1 Wolfe, "'Me' Decade."

2 In 2010, the *New York Post* wrote, "The South Bronx (along with Brooklyn's Brownsville, Bushwick, and Bedford-Stuyvesant neighborhood, and Manhattan's Harlem and Lower East Side) was indeed burning. Seven different census tracts in The Bronx lost more than 97% of their buildings to fire and abandonment between 1970 and 1980; 44 tracts (out of 289 in the borough) lost more than 50%. 'The smell is one thing I remember,' says retired Bronx firefighter Tom Henderson. 'That smell of burning—it was always there, through the whole borough almost.'" Flood, "Why the Bronx Burned."

3 Brink-Johnson and Lubin, "Structural Racism," 6.

4 Mahler, *Ladies and Gentlemen*, 6.

5 Mehlman, "Elevating the Urban Environment," 42.

6 Joan Aho, "Information for the Press," Public Relations Department, Citicorp Center, revised January 1978, quoted in Hartley, "Implications," 265.

7 Hartley, "Implications," 264–65.

8 Hartley, "Implications," 265.

9 Hugh Stubbins to Henry J. Muller, November 2, 1970, Papers of Hugh Stubbins.

10 Hugh Stubbins and Associates, "FNCB Alternative Studies," February 7, 1972, Papers of Hugh Stubbins.

11 Hugh Stubbins and Associates, "FNCB Alternative Studies."

12 Hugh Stubbins to Richard Roth Sr., June 29, 1972, Papers of Hugh Stubbins.

13 "Project SERENE," November 20, 1978, 2, Papers of William LeMessurier.

14 Dryansky, "Hugh's Masterpiece," 206.

15 "Project SERENE," 28.

16 LeMessurier, "Fifty-Nine-Story Crisis."

17 LeMessurier, "Fifty-Nine-Story Crisis."

18 "Project SERENE," 3.

19 Hugh Stubbins to John S. Reed, February 15, 1979, Papers of Hugh Stubbins.

20 "Legs Centered under Each Face," 69.

21 Stubbins, *Architecture*, 27.

22 Von Eckardt, "Wolf Von Eckardt on Architecture," 27.

23 Landmarks Preservation Commission, "Citicorp Center," 7, quoting Gardiner, "Transparently Beautiful," 32.

24 Hugh Stubbins to Henry DeFord III, June 27, 1973, Papers of Hugh Stubbins.

25 Zweig, *Wriston*, 544.

26 Dryansky, "Hugh's Masterpiece," 204.

27 Zweig, *Wriston*, 603.

28 "Plan for Skyscraper on Lexington Ave.," 47.

29 Mehlman, "Elevating the Urban Environment," 42.

30 Landmarks Preservation Commission, "Citicorp Center," 12, quoting Goldberger, "No Taint of Materialism."

31 Landmarks Preservation Commission, "Citicorp Center," quoting Citicorp press release, 1977, 5, Citi Center for Culture.

32 For a description of the sunken plaza, see Landmarks Preservation Commission, "Citicorp Center," 10.

33 "Citicorp in the 21st Century," unidentified newspaper article found in Papers of Hugh Stubbins.

34 Remarks of Hugh Stubbins at the International Conference on the Use of Church Properties, May 13–14, 1975, Papers of Hugh Stubbins.

35 Mehlman, "Elevating the Urban Environment," 48.

36 LeMessurier, "Fifty-Nine-Story Crisis."

37 Von Eckardt "Wolf Von Eckardt on Architecture," 26.

38 Schmertz, "Citicorp Center," 115.

39 Von Eckardt "Wolf Von Eckardt on Architecture," 28.

40 Huxtable, "New Urban Image?," 21.

41 According to Nusbaum's obituary in *Engineering News-Record*, HRH "cut ties with Trump after the developer's widely-touted 1986 renovation of the city-owned Wollman ice skating rink in Central Park, which Nusbaum said HRH agreed to do for cost but whose work Trump never acknowledged." Rubin, "NYC Builder Who Challenged Trump."

42 Oser, "About Real Estate," 73.

43 Morgenstern, interview with Robertson, 1.

44 Oser, "About Real Estate," 73.

45 "Legs Centered under Each Face," 69.

46 Schmertz, "Citicorp Center," 116; Mehlman, "Elevating the Urban Environment," 43.

47 "Project SERENE," 28.

48 Morgenstern, "Fifty-Nine-Story Crisis," 46.

49 LeMessurier, "Fifty-Nine-Story Crisis."

50 Easley Hamner later expressed some misgivings about the decision to hide the chevrons. "Perhaps if we were doing [it] again and realized the elegance of the frame we would have accentuated the structure more," he told reporters. Horsley, "New Wrinkle," 262.

51 Goldberger, "Citicorp's Center Reflects Synthesis," 57.

52 Horsley, "Citicorp's Skyscraper 'Weathering,'" R1; Stubbins, *Architecture*, 24.

53 Goldberger, "Citicorp's Center Reflects Synthesis," 57.

54 "Legs Centered under Each Face," 68.

55 Landmarks Preservation Commission, "Citicorp Center," 11, quoting press release, July 24, 1973, Citi Center for Culture.

56 Stubbins, *Architecture*, 26–27.

57 Hartley, "Implications," 221.

58 "Legs Centered under Each Face," 68.

59 "Legs Centered under Each Face," 69; LeMessurier, "Fifty-Nine-Story Crisis."

60 LeMessurier, "Fifty-Nine-Story Crisis."

61 Spielvogel v. Lynch, 127 A.D.2nd 532, 532 (N.Y. App. Div. 1987).

62 Hartley, "Implications," 129.

63 Higgins, "Keeping Skyscrapers."

64 Hartley, "Implications," 133. LeMessurier determined that deflection had to be controlled to an angle of 1/400 under an average force of thirty pounds per square foot. "Project SERENE," 4.

65 See Rippeteau, "How Much Wind Can a Building Take?," R1.

66 S. Roberts, *Wind Wizard*.

67 LeMessurier, "Fifty-Nine-Story Crisis."

68 LeMessurier, "Fifty-Nine-Story Crisis."

69 LeMessurier, "Fifty-Nine-Story Crisis."

70 Dryansky, "Hugh's Masterpiece," 237.

71 Mehlman, "Elevating the Urban Environment," 46.

72 Hartley, "Implications," 137.

73 Hugh Stubbins to Robert Dexter, July 1, 1974, Papers of Hugh Stubbins.

74 LeMessurier recalled, "[We were] dressed to the nines in pin stripe suits for [the] meeting and in walks a young man with nubby sweater on and no tie and kind of wrinkled pants, I thought he was going to bring the coffee or something and it turned out to be John Reed. He was really laid back, but very approachable, serious. . . . He didn't make an immediate decision, said he'd have to get a second opinion, and knew where to go, because at MIT he'd roomed with a man studying structural dynamics." Morgenstern, interview with LeMessurier, 9.

75 Hartley, "Implications," 139.

76 Per LeMessurier, "Mr. Robert McNamara was in general charge of LeMessurier Associates' role in the Joint Venture from its formation until the spring of 1975. On January 1, 1974 Mr. McNamara was advanced in status from Associate to Principal and Vice President of LeMessurier Associates/SCI (LeMessurier Associates, Inc. officially became Sippican Consultants International, Inc. after a merger on March 31, 1973). By the spring of 1975, the construction of Citicorp Center was well along, and McNamara was assigned to other duties and moved to Rome, Italy. Mr. Stanley Goldstein assumed McNamara's responsibilities." "Project SERENE," 6.

77 Robert J. McNamara, telephone conversation with the author, June 15, 2023.

78 "Engineering For Architecture," 69.

CHAPTER 3. "A SKYSCRAPER FOR THE PEOPLE"

1 Oser, "About Real Estate," 73.

2 Horsley, "New Wrinkle," 262.

3 Walter B. Wriston, "Remarks at the Topping Out Ceremony," October 6, 1976, Location: MS134.001.003.00010, Wriston Papers.

4 "Executives and Workmen Celebrate," 21.

5 Walter B. Wriston, "Remarks at the Citicorp Cornerstone Ceremony," October 31, 1976, Location: MS134.001.003.00011, Wriston Papers.

6 Landmarks Preservation Commission, "Citicorp Center," 7, quoting Stubbins, "Skyscraper for People," 27.

7 Morgenstern, interview with LeMessurier, 3.

8 Huxtable, "New Urban Image?," 21.

9 Reel, "East Side, West Side," 165.

10 Mehlman, "Elevating the Urban Environment," 48.

11 Goldberger, "Citicorp's Center Reflects Synthesis," 57.

12 Campbell, "Citicorp Center," A10.

13 Egan, "Citicorp Center," E1.

14 Landmarks Preservation Commission, "Citicorp Center," quoting Gardiner, "Transparently Beautiful."

15 Goldberger, "Diversity Marks Architecture Awards," 21.

16 *Modern Steel Construction* 19, nos. 1–2 (1979): 14.

17 Landmarks Preservation Commission, "Citicorp Center," 13, quoting American Institute of Architects, *AIA Guide to New York City*, 152.

18 Landmarks Preservation Commission, "Citicorp Center," 19.

19 Morgenstern, interview with LeMessurier, 1.

20 LeMessurier, "Fifty-Nine-Story Crisis."

21 LeMessurier, "Fifty-Nine-Story Crisis."

CHAPTER 4. DIANE HARTLEY

1 Fitzgerald, *This Side of Paradise*, 42.

2 Fitzgerald, *This Side of Paradise*, 53, 54.

3 Fitzgerald, *This Side of Paradise*, 42, 36.

4 Much of the information contained in this chapter comes from various telephone conversations and email exchanges between the author and Diane Hartley; Ley, "Student Saves Skyscraper"; and a presentation given by Diane Hartley at the Washington, DC office of Simpson Gumpertz & Heger (structural engineers) on October 14, 2019, a recorded copy of which Hartley provided to the author.

5 Ley, "Student Saves Skyscraper."

6 J. Sullivan, "David Billington."

7 J. Sullivan, "David Billington."

8 Ley, "Student Saves Skyscraper."

9 Ley, "Student Saves Skyscraper."

10 "People to Know Delayed Reaction," 66.

11 Billington, *Tower and the Bridge*, 5; Gauvreau, "Current Realities," 234.

12 Billington, *Tower and the Bridge*, 5.

13 Billington, *Tower and the Bridge*, 16, 17.

14 Santosuosso, "Swivel-Hipped Prophet," 1.

15 Hartley, "Implications," 1.

16 Citicorp had not yet transitioned its offices to the new tower.

17 Ley, "Student Saves Skyscraper."

18 Hartley, "Implications," appendix B. In determining static wind loading, Hartley used the National Association of Architectural Metal Manufactures Standard WL-10-67, which set the maximum annual extreme-mile velocity in New York City at eighty miles per hour at thirty feet above the ground. "This is the velocity used in static wind-resistant design of structures in this area," she wrote. Hartley, "Implications," 365.

19 "Project SERENE," November 20, 1978, 5, Papers of William LeMessurier. LeMessurier acknowledged that the code required consideration of wind "from every direction" but insisted that the industry interpreted that to mean winds from the north, south, east, and west—not on the diagonal. Morgenstern, interview with LeMessurier, 11.

20 Hartley, "Implications," 109.

21 Hartley, "Implications," 376.

22 Hartley, "Implications," 377.

23 Ley, "Student Saves Skyscraper."

24 Gottlieb, "She Draws the Interest," 155.

25 Hartley, presentation at the offices of Simpson Gumpertz & Heger.

26 Diane Hartley, telephone conversation with author, May 4, 2023.

27 Hartley, "Implications," 377.

28 Regan, "Structural Integrity Video."

29 "Legs Centered under Each Face," 69.

30 Kremer, "(Re)Examining the Citicorp Case," 321, referring to Robert McNamara to Eugene Kremer, email, February 11, 2002.

31 Billington's handwritten list of comments on Hartley's thesis were provided to the author by Hartley.

32 "Legs Centered under Each Face," 69.

CHAPTER 5. THE MYSTERY STUDENT

1 Dryansky, "Hugh's Masterpiece," 204.

2 "Bank Reached for Topless Neighbors," 5; Zweig, *Wriston*, 603.

3 "Project SERENE," November 20, 1978, 7, Papers of William LeMessurier.

4 Zweig, *Wriston*, 603.

5 "Project SERENE," 8.

6 Morgenstern, "Fifty-Nine Story Crisis," 45.

7 Morgenstern, "Fifty-Nine Story Crisis," 46.

8 Morgenstern, interview with LeMessurier, 3.

9   Lee DeCarolis to Peggy Walsh, email, November 2, 2011, provided to the author by DeCarolis, August 8, 2023.
10 DeCarolis, "Citicorp Building."
11 Regan, "Structural Integrity Video."

CHAPTER 6. "SOME VERY PECULIAR BEHAVIOR"

1   Kremer, "(Re)Examining the Citicorp Case," 321, referring to Robert McNamara to Eugene Kremer, email, February 11, 2002; and Robert McNamara to Joe Morgenstern, December 18, 1995.
2   LeMessurier, "Fifty-Nine-Story Crisis."
3   "Project SERENE," November 20, 1978, 8, Papers of William LeMessurier.
4   Morgenstern, interview with LeMessurier, 10.
5   Morgenstern, "Fifty-Nine Story Crisis," 46.
6   LeMessurier, "Fifty-Nine-Story Crisis."
7   LeMessurier, "Fifty-Nine-Story Crisis."
8   Morgenstern, interview with LeMessurier, 10.
9   "It is interesting to note that, in the contract documents, the connections for these diagonals were shown to be by complete penetration weld. Had these details been followed, the connections would have been stronger than the members themselves—and very substantially stronger than was required." Skilling, Helle, Christiansen, Robertson, "Structural Investigation of Citicorp Center," July 5, 1979, 13, Papers of William LeMessurier.
10 LeMessurier, "Fifty-Nine-Story Crisis." LeMessurier also relied on a statement by Murray Shapiro from the Office of James Ruderman, recounting a conversation with one of the authors of the New York City Building Code, who told him that winds of "any direction," as used in the code, was meant by the code committee to imply only "winds perpendicular to any of the four faces of a square or rectangular building." "Project SERENE," 30.
11 Morgenstern, "Fifty-Nine Story Crisis," 46.
12 Story and quotes from Chass, "Martin Resigns," 1.
13 Morgenstern, interview with LeMessurier, 11.
14 Morgenstern, interview with LeMessurier, 11.
15 "Project SERENE," 28.
16 Morgenstern, "Fifty-Nine Story Crisis," 46.
17 "Project SERENE," 9. LeMessurier later said that his New York engineering team had interpreted the chevron members of the bracing system as "trusses" rather than "columns" pursuant to the American Institute of Steel Construction Specifications, thus exempting them from the safety factor set forth in the code. See LeMessurier, "Fifty-Nine-Story Crisis."
18 "Project SERENE," 9. See also Morgenstern, interview with LeMessurier, 13.

19 "Project SERENE," 10.

20 LeMessurier, "Fifty-Nine-Story Crisis."

CHAPTER 7. "VERTICAL URBANISM"

1 "Sally Regenhard."

2 Much of Sally Regenhard's story and her quotes are found in Duke, "From Anger to Action."

3 Duke, "From Anger to Action."

4 Duke, "From Anger to Action."

5 Duke, "From Anger to Action."

6 Regenhard, statement to the National Commission on Terrorist Attacks upon the United States.

7 Duke, "From Anger to Action."

8 See National Institute of Standards and Technology, "FAQs—NIST WTC Towers Investigation."

9 See Banovic et al., "Role of Metallurgy."

10 Clinton, "Statement in Response."

11 Morris, "Tips for Better, Safer Buildings."

12 "When designing buildings one can only anticipate the worst-case scenario known at the time of construction," wrote Oral Buyukozturk, professor of civil and environmental engineering at MIT. "If this scenario is known, collapse can be prevented by innovative engineering design of materials and structures. The WTC towers were indeed designed to withstand the impact of a large commercial aircraft. They were not, however, designed to withstand the prolonged effect of fire resulting from a bomb in the guise of a fully fueled aircraft." Buyukozturk, "How Safe Are Our Skyscrapers?"

13 One research engineer with NIST called the New York City code's failure to require analysis of quartering winds "dangerously negligent." Duthinh, "Blown Away," 3.

14 Kremer, "(Re)Examining the Citicorp Case," 319.

15 Kremer, "(Re)Examining the Citicorp Case."

16 Kremer, "(Re)Examining the Citicorp Case," 317, quoting Matthys Levy to Eugene Kremer, email, May 29, 2002.

17 Skilling, Helle, Christiansen, Robertson, "Structural Investigation of Citicorp Center," July 5, 1979, 12, Papers of William LeMessurier.

18 "Project SERENE" (early version, n.d.), 5, Papers of William LeMessurier. On this point, he told the journalist Joe Morgenstern, "I had to confess that I knew better, because I had written the building code in Massachusetts myself, and when it came to this matter I said if it's a rectangular building you must analyze for it, in effect, a quartering wind. But anyway . . . the Massachusetts code was never widely published or known, and anyway these computations were made in New York by New Yorkers, and they didn't have that in the code, so I

can't blame the guys in my office." Morgenstern, interview with LeMessurier, 12.

19 Kremer, "(Re)Examining the Citicorp Case," 319, quoting Robert McNamara to Eugene Kremer, email, February 11, 2002. LeMessurier wrote, "the effect of such a wind on the columns is *not* critical." "Project SERENE," 28.

20 "Project SERENE," 5.

21 "Project SERENE," 28.

22 "Project SERENE," 5.

23 "Project SERENE," 3.

24 Korman, "Critics Grade Citicorp Confession," 10.

25 S. Roberts, *Wind Wizard*, 119–20.

26 Duthinh, "Blown Away," 4.

27 Robert J. McNamara, telephone conversation with the author, June 15, 2023.

28 Buyukozturk, "How Safe Are Our Skyscrapers?," 2.

29 Council on Tall Buildings and Urban Habitat, "Global Impact of 9/11 on Tall Buildings."

30 Council on Tall Buildings and Urban Habitat, "Global Impact of 9/11 on Tall Buildings."

31 Council on Tall Buildings and Urban Habitat, "Global Impact of 9/11 on Tall Buildings."

32 Council on Tall Buildings and Urban Habitat, "Global Impact of 9/11 on Tall Buildings."

33 Piber, "Megacities vs. Urban Sprawl," 22.

34 Piber, "Megacities vs. Urban Sprawl," 22.

35 Kamin, "City Building," 55.

CHAPTER 8. ONE IN SIXTEEN

1 Morgenstern, interview with LeMessurier, 17.

2 Morgenstern, interview with LeMessurier, 17.

3 "Probabilities are . . . unique for each building, because a building may be sensitive to a northwest wind but not a southwest wind," LeMessurier later said. "So just to say that a wind blew in New York of a certain velocity doesn't mean anything. Did it blow from a certain direction to be critical for this building, especially since we're talking about wind blowing from the diagonals?" Morgenstern, interview with LeMessurier, 15.

4 Morgenstern, interview with LeMessurier, 27, 13.

5 BBC, "All Fall Down."

6 Morgenstern, "Fifty-Nine Story Crisis," 47; Morgenstern, interview with LeMessurier, 14.

7 Morgenstern, interview with LeMessurier, 14.

8 BBC, "All Fall Down."

9 LeMessurier's friend and business partner William Thoen drove him to Maine. "When Bill first became aware of the problem, the depths of which insidiously unfolded upon him, I drove him from Cambridge to his island in Maine (he was too otherwise concerned to drive himself)," recalled Thoen. Bill Thoen, to the author, email, December 18, 2023.

10 LeMessurier, "Fifty-Nine-Story Crisis."

11 Morgenstern, "Fifty-Nine Story Crisis," 47.

CHAPTER 9. "A CATALOGUE OF FAILURES"

1 "Project SERENE," November 20, 1978, 31, Papers of William LeMessurier.

2 Petroski, *To Forgive Design*, 39. Petroski wrote, "Even failures that occurred years, decades, centuries, and millennia ago can inform our designs today and prevent them from turning into failures." Petroski, *To Forgive Design*, 37.

3 Tacitus, *Complete Works of Tacitus*, 180–81 (4.62).

4 Davis, *State of Washington*, 34.

5 Washington State Department of Transportation, "Tacoma Narrows Bridge History—Eyewitness Accounts."

6 Washington State Department of Transportation, "Tacoma Narrows Bridge History—Eyewitness Accounts."

7 Washington State Department of Transportation, "Tacoma Narrows Bridge History—Lessons."

8 Ammann, "Present Status of Design," 253.

9 Witcher, "From Disaster to Prevention," 45.

10 West Virginia Department of Transportation, "Silver Bridge."

11 Witcher, "From Disaster to Prevention," 45.

12 Witcher, "From Disaster to Prevention," 46.

13 Petroski, *To Forgive Design*, 160.

14 West Virginia Department of Transportation, "Silver Bridge."

15 Petroski, *To Forgive Design*, 167, quoting National Transportation Safety Board, *Highway Accident Report*, 126.

16 Petroski, *To Forgive Design*, 168.

17 Petroski, *To Forgive Design*, 222.

18 Petroski, *To Forgive Design*, 340–41.

19 *Authentic History of the Lawrence Calamity*, 8.

20 *Authentic History of the Lawrence Calamity*, 22.

21 *Authentic History of the Lawrence Calamity*, 41–42.

22 Walsh and Hudson, "Tea Dance."

23 Walsh and Hudson, "Tea Dance."

24 Duncan v. Missouri Bd. for Architects, 744 S.W.2d 524, 524 (Mo. App. E.D., 1988).

25 Petroski, *To Engineer Is Human*, 85.

26 National Bureau of Standards, *Investigation of the Kansas City Hyatt*, vi.

27 National Bureau of Standards, *Investigation of the Kansas City Hyatt*, iii.

28 *Duncan*, 744 S.W.2d at 540.

29 Rechtien, "Hyatt Regency Disaster Revisited," 60.

30 National Society of Professional Engineers, "Code of Ethics for Engineers."

31 Petroski, *To Engineer Is Human*, 52.

32 See Jiang et al., "Review on Quantitative Measures," 129.

CHAPTER 10. "A SERIOUS AND DEADLY MATTER"

1 Morgenstern, interview with LeMessurier, 1.

2 Morgenstern, interview with LeMessurier, 1.

3 Morgenstern, interview with LeMessurier, 15.

4 Morgenstern, "Fifty-Nine-Story Crisis," 48.

5 Morgenstern, interview with LeMessurier, 16.

6 Morgenstern, interview with LeMessurier, 16.

7 "Outer Banks Waterspout Leaves 1 Dead," 1.

8 Al Romaneski, Memorandum to Record, July 31, 1978, Papers of William LeMessurier.

9 Al Romaneski to Roy Vince, Shand, Morahan & Co., Inc., July 31, 1978, Papers of William LeMessurier.

CHAPTER 11. REVELATIONS

1 LeMessurier, "Fifty-Nine-Story Crisis."

2 Morgenstern, "Fifty-Nine-Story Crisis," 49.

3 Skilling, Helle, Christiansen, Robertson, "Structural Investigation of Citicorp Center," July 5, 1979, 1, Papers of William LeMessurier.

4 LeMessurier, "Fifty-Nine-Story Crisis."

5 Post, "World-Renowned Structural Engineer," 16.

6 Hickman, "Leslie E. Robertson."

7 Post, "World-Renowned Structural Engineer," 16.

8 Stern, Gilmartin, and Mellins, *New York 1930*, 530.

9 "Office Building Here to Be Largest Yet," 1.

10 LeMessurier, "Fifty-Nine-Story Crisis."

11 See Morgenstern, "Fifty-Nine-Story Crisis," 48.

12 Skilling, Helle, Christiansen, Robertson, "Structural Investigation," 12.

13 Morgenstern, interview with Robertson, 3; Morgenstern, "Fifty-Nine-Story Crisis," 49.

14 Morgenstern, interview with LeMessurier, 18.

15 Morgenstern, interview with LeMessurier, 18, 21.

16 Morgenstern, interview with Robertson, 4. Though Robertson took pride in the work that he, LeMessurier, and others would accomplish at the Citicorp Building, he looked back at the experience with some disdain. "I hate the place," he told the journalist Joe Morgenstern in a 1993 interview. "It was a terrible impingement on my life. I have other goals, and

buildings should not need to be repaired. . . . I really don't like Citicorp Center, it's one of those places I don't go to." Morgenstern, interview with Robertson, 4.

17 Morgenstern, interview with LeMessurier, 18.
18 Morgenstern, interview with Robertson, 3.
19 BBC, "All Fall Down."
20 BBC, "All Fall Down."
21 LeMessurier, "Fifty-Nine-Story Crisis."
22 LeMessurier, "Fifty-Nine-Story Crisis."
23 Reed became chairman and CEO of Citicorp in September 1984. "Wriston may have wanted a visionary genius to succeed him, but he was stuck with me," joked Reed in 2019. Lee, "Bankers That Define the Decades." He would remain with Citicorp for thirty-five years, and in 2003, he became chairman of the New York Stock Exchange.
24 Morgenstern, interview with LeMessurier, 19.
25 Zweig, *Wriston*, 1.
26 Morgenstern, interview with LeMessurier, 19.
27 Zweig, *Wriston*, jacket flap.
28 Zweig, *Wriston*, 1.
29 P. Sullivan, "Walter B. Wriston."
30 Zweig, *Wriston*, 2.
31 P. Sullivan, "Walter B. Wriston."
32 Morgenstern, interview with Leslie Robertson, 6.
33 Morgenstern, interview with Leslie Robertson, 6.
34 Dwyer, "Didja Hear about Citicorp Tower?"
35 Morgenstern, "Fifty-Nine-Story Crisis," 50.
36 BBC, "All Fall Down."
37 Morgenstern, interview with LeMessurier, 21.
38 "Project SERENE," November 20, 1978, 15, Papers of William LeMessurier.

CHAPTER 12. MOBILIZATION

1 Waldorf Astoria, "History."
2 Morgenstern, interview with LeMessurier, 26.
3 Morgenstern, interview with LeMessurier, 26.
4 BBC, "All Fall Down"
5 Morgenstern, interview with LeMessurier, 21.
6 Morgenstern, interview with LeMessurier, 21.
7 "Project SERENE," November 20, 1978, 16, Papers of William LeMessurier.
8 Confidential Memorandum of Meeting—Weather Forecasting—Project SERENE, August 4, 1978, Papers of William LeMessurier.
9 Contract between Leslie Robertson and Irving A. Singer, August 4, 1978, Papers of William LeMessurier.

10 Robertson was sadly proven correct during the attack on New York's World Trade Center in February 1993. He would point out, "We have emergency generators at the World Trade Center that lasted for fifteen minutes. . . . So that's the problem with that kind of device, you really have to look at the overall system's reliability." Morgenstern, interview with Robertson, 4.

CHAPTER 13. "ANGEL OF THE BATTLEFIELD"

1 Internal handwritten notes in LeMessurier's papers housed at the Harvard Graduate School of Design indicate that the group considered the following historical events in developing "Contingency Planning Activities":
—Central America Earthquake Disaster Plan (1976)
—Power Blackout Business Continuity Planning (1977)
—Emergency Transit Strike Operations (During 1966 implementation resulted in facilitating effective movement of 9190 of staff)
—Domestic/International Kidnapping + Extortion
—Crises Management Planning For Terrorist Activity
—NYPD Strike Threat Planning Committee
Evacuation Planning, August 7, 1978, Contingency Planning Activities, Papers of William LeMessurier.
2 Corwin and Miles, "Impact Assessment," 13.
3 Corwin and Miles, "Impact Assessment," 14.
4 Martinez, "Transit Strike."
5 Martinez, "Transit Strike."
6 Evacuation Planning.
7 Olson and Olson, "Guatemala Earthquake," 69.
8 McGee, opening statement, 2.
9 Olson and Olson, "Guatemala Earthquake," 75.
10 Olson and Olson, "Guatemala Earthquake," 76.
11 Olson and Olson, "Guatemala Earthquake," 77. Senator Gale W. McGee said, "The U.S. response to the disaster and our handling of the relief operations should serve as a case study for future operations of this scope. I am advised that the coordination of the disaster relief operations was exceptional. The government of Guatemala responded quickly and worked in close concert with the United States. The mobilization of the military and civilian sectors of the U.S. government was outstanding for the precision in which damage assessments were made, in the determination of immediate and most pressing needs of the population affected by the earthquake, and in the logistics required for transporting vitally needed supplies to the hard hit rural areas." McGee, opening statement, 4.
12 Skilling, Helle, Christiansen, Robertson, "Structural Investigation of Citicorp Center," July 5, 1979, 7, Papers of William LeMessurier.
13 LeMessurier, "Fifty-Nine-Story Crisis."

14 Morgenstern, interview with Robertson, 5.

15 R. W. Clarke to John Curtus, August 8, 1978, Papers of William LeMessurier.

16 Clarke to Curtus.

17 Morgenstern, "Fifty-Nine-Story Crisis," 51.

18 Hebert, *Preliminary Report*, 1.

19 Information about Clara Barton and the history of the American Red Cross is found at Red Cross, "Founder Clara Barton"; and Red Cross, "Brief History of the American Red Cross."

20 Red Cross, "Founder Clara Barton."

21 Morgenstern, interview with Reilly, 1.

22 Morgenstern, interview with Reilly, 1.

CHAPTER 14. "WHITE LIES"

1 Kleiman, "New York Cheers," B2; "Transcript of Carter's Talk at City Hall," B2.

2 "Transcript of Carter's Talk at City Hall," B2.

3 "Transcript of Carter's Talk at City Hall," B2.

4 "Transcript of Carter's Talk at City Hall," B2.

5 Kleiman, "New York Cheers," B2.

6 Morgenstern, interview with LeMessurier, 20.

7 LeMessurier, "Fifty-Nine-Story Crisis."

8 Morgenstern, interview with LeMessurier, 20.

9 Skilling, Helle, Christiansen, Robertson, "Structural Investigation of Citicorp Center," July 5, 1979, 2, Papers of William LeMessurier.

10 Final Draft—August 4, 1978, W. J. LeMessurier's Statement, Papers of William LeMessurier.

11 LeMessurier, "Fifty-Nine-Story Crisis."

12 LeMessurier, "Fifty-Nine-Story Crisis."

13 Morgenstern, interview with LeMessurier, 23.

14 "Project SERENE," November 20, 1978, 18, Papers of William LeMessurier.

15 Morgenstern, interview with LeMessurier, 24.

16 Skilling, Helle, Christiansen, Robertson, "Structural Investigation," 38.

17 Public Affairs Department Press Release, Lamson B. Smith, Papers of William LeMessurier.

18 Morgenstern, interview with Robertson, 2.

19 Morgenstern, interview with Robertson, 3. Robertson was apparently not as circumspect after the fact. LeMessurier grumbled that Robertson talked about the story, mostly to other design professionals in the New York area. Morgenstern, interview with LeMessurier, 32. Mike Reilly of the American Red Cross said that "Leslie Robertson seemed to make a career out of that building. . . . Every now and then he'd pop up in the media." Morgenstern, interview with Reilly, 1.

20 "Citicorp Bldg. to Get 1M Wind Bracing," 10. It is doubtful that this information came from LeMessurier, who insisted that actual wind velocity was not relevant to quantifying building strength. "Probabilities are all that you can really talk about, and they're unique for each building," he later said. Morgenstern, interview with LeMessurier, 15.

21 "Citicorp Bldg. to Get 1M Wind Bracing," 10.

22 "Citicorp Bldg. to Get 1M Wind Bracing," 10.

23 "Citicorp to Brace for Big Wind," 26.

24 BBC, "All Fall Down."

CHAPTER 15. "A THOUSAND-YEAR WIND"

1 Clark, *Hurricane Cora*, 1. The first storm to receive hurricane status based on satellite imaging only was Hurricane Doris in 1975. Clark, *Hurricane Cora*, 2.

2 United Press International, "Citicorp to Brace for Big Wind," 26.

3 Morgenstern, interview with LeMessurier, 24.

4 "Engineer's Afterthought," 11.

5 Morgenstern, interview with LeMessurier, 24.

6 "Engineer's Afterthought," 11.

7 Morgenstern, interview with LeMessurier, 24.

8 LeMessurier, "Fifty-Nine-Story Crisis."

9 BBC, "All Fall Down."

10 LeMessurier, "Fifty-Nine-Story Crisis."

11 Friendly, "Story of the Newspaper Walkout," 86.

12 Stetson, "Talks Intensify," 6.

13 Friendly, "Story of the Newspaper Walkout," 86.

14 Stetson, "Talks Intensify," 6.

15 Edes, "Was a Newspaper Strike a Bigger Villain?"

16 "Project SERENE," November 20, 1978, 20, Papers of William LeMessurier.

17 "Project SERENE," 21.

18 Morgenstern, interview with LeMessurier, 26.

19 LeMessurier wrote, "Throughout the repair period, the principal concern of the Bank in addition to bringing the building to a safe state as soon as possible was to cause the least disturbance to tenants." "Project SERENE," 25.

20 Morgenstern, interview with LeMessurier, 26.

21 "Project SERENE," 20.

22 Skilling, Helle, Christiansen, Robertson, "Structural Investigation of Citicorp Center," July 5, 1979, 13, Papers of William LeMessurier.

23 Skilling, Helle, Christiansen, Robertson, "Structural Investigation," 13; "Engineer's Afterthought," 11.

24 Skilling, Helle, Christiansen, Robertson, "Structural Investigation," 35.

25 BBC, "All Fall Down."

CHAPTER 16. SAFETY, SECRECY, AND SURVEILLANCE

1 Morgenstern, interview with LeMessurier, 27.
2 Morgenstern, interview with LeMessurier, 25. "The reason I was there," LeMessurier said, "was because I was on top of fixing the building, and . . . I was the one who understood the building better than anyone alive, and I knew what had to be fixed first." Morgenstern, interview with LeMessurier, 25.
3 Morgenstern, interview with Reilly, 1.
4 BBC, "All Fall Down."
5 Morgenstern, interview with LeMessurier, 26.
6 Morgenstern, interview with LeMessurier, 15.
7 Morgenstern, interview with LeMessurier, 25.
8 Skilling, Helle, Christiansen, Robertson, "Structural Investigation of Citicorp Center," July 5, 1979, 7, Papers of William LeMessurier.
9 Morgenstern, interview with LeMessurier, 25.
10 Skilling, Helle, Christiansen, Robertson, "Structural Investigation," 7. LeMessurier harbored resentment toward Robertson even about Kinemetrics. "I'm not sure what the purpose [of the building monitors] was, except to educate Robertson, which was nice," LeMessurier later said acerbically. Morgenstern, interview with LeMessurier, 25.
11 Morgenstern, "Fifty-Nine-Story Crisis," 52.

CHAPTER 17. A CRITICAL ANALYSIS

1 Richard Henige to the author, email, December 28, 2023.
2 Robertson later wrote, "While it could be argued that these LeMessurier people might not be completely unbiased in this investigation, experience has now shown that they pursued their work with competence and diligence, and we have every confidence in their integrity." Skilling, Helle, Christiansen, Robertson, "Structural Investigation of Citicorp Center," July 5, 1979, 8, Papers of William LeMessurier.
3 Albert L. Romaneski to Harold Moskowitz, May 2, 1979, Papers of William LeMessurier.
4 Morgenstern, "Fifty-Nine-Story Crisis," 52.
5 Much of the information regarding the Citicorp design review was provided by former LeMessurier engineer Richard Henige. Henige to the author, email, December 28, 2023.
6 Skilling, Helle, Christiansen, Robertson, "Structural Investigation," 19.
7 Skilling, Helle, Christiansen, Robertson, "Structural Investigation," 19.
8 Skilling, Helle, Christiansen, Robertson, "Structural Investigation," 19.
9 "Project SERENE," November 20, 1978, 23, Papers of William LeMessurier.
10 "Project SERENE," 24–25.
11 Skilling, Helle, Christiansen, Robertson, "Structural Investigation," 23.
12 "Project SERENE," 25.

13 Skilling, Helle, Christiansen, Robertson, "Structural Investigation," 23.

14 Morgenstern, interview with LeMessurier, 31, 27.

CHAPTER 18. ELLA

1 Larson, *Isaac's Storm*, 25.

2 For a detailed explanation of the butterfly effect and chaos theory, see Lorenz, *Essence of Chaos*.

3 Morgenstern, interview with LeMessurier, 32.

4 Lawrence, *Preliminary Report*, 1.

5 Morgenstern, interview with LeMessurier, 29.

6 National Weather Service, "Tropical Storm Ella Special Advisory Number 1."

7 Morgenstern, interview with LeMessurier, 29.

8 National Weather Service, "Special Request."

9 Morgenstern, "Fifty-Nine-Story Crisis," 53.

10 Morgenstern, "Fifty-Nine-Story Crisis," 53.

CHAPTER 19. "EVENTS NOBODY ENVISIONED"

1 Roberts, "Roxcy Bolton."

2 The review team further identified the need for bracing of the core columns on the nineteenth and twenty-seventh floors, reinforcement of the transfer trusses on the ninth and fourteenth floors, completion of a ninth-floor diaphragm wind analysis, and issues with the core bracing on the subcellar of the fourteenth floor. Skilling, Helle, Christiansen, Robertson, "Structural Investigation of Citicorp Center," July 5, 1979, 8, Papers of William LeMessurier.

3 Hans H. Angermueller to Hugh Stubbins and Associates, Emery Roth & Sons, LeMessurier Associates, and James Ruderman, September 13, 1978, Papers of William LeMessurier.

4 LeMessurier, "Fifty-Nine-Story Crisis."

5 Al Romaneski, Memorandum to Record, August 22, 1978, Papers of William LeMessurier.

6 Romaneski, Memorandum to Record, August 22, 1978.

7 Al Romaneski, Memorandum to Record, September 25, 1978, Papers of William LeMessurier.

8 Skilling, Helle, Christiansen, Robertson, "Structural Investigation," 37.

9 "Project SERENE," November 20, 1978, 28, Papers of William LeMessurier.

10 "Project SERENE," 9.

11 "Project SERENE," 11.

12 "Project SERENE," 29.

13 "Project SERENE," 29.

14 "Project SERENE," 30.

15 "Project SERENE," 30.

16 "Project SERENE," 31.

17 "Project SERENE," 31.

18 "Project SERENE," 33.

19 "Project SERENE," 32–33.

20 "Project SERENE," 34.

21 "Project SERENE," 27.

22 Hans H. Angermueller to Hugh Stubbins and Associates, Emery Roth & Sons, LeMessurier Associates, and James Ruderman, May 3, 1979, Papers of William LeMessurier.

23 Skilling, Helle, Christiansen, Robertson, "Structural Investigation," 4.

24 Skilling, Helle, Christiansen, Robertson, "Structural Investigation," 8.

25 Morgenstern, interview with Robertson, 5.

26 Skilling, Helle, Christiansen, Robertson, "Structural Investigation," 38.

27 Skilling, Helle, Christiansen, Robertson, "Structural Investigation," 38.

28 Skilling, Helle, Christiansen, Robertson, "Structural Investigation," 37.

29 Robert H. Dexter to William LeMessurier, July 25, 1979, Papers of William LeMessurier.

30 William J. LeMessurier to Carl M. Sapers, July 25, 1980, Papers of William LeMessurier.

31 LeMessurier, "Fifty-Nine-Story Crisis."

32 Morgenstern, interview with LeMessurier, 35; Morgenstern, "Fifty-Nine-Story Crisis," 53.

33 Morgenstern, "Fifty-Nine-Story Crisis," 53.

34 Morgenstern, interview with Robertson, 5.

35 Morgenstern, interview with LeMessurier, 35.

EPILOGUE

1 The full account of Joe Morgenstern's discovery of the story, contact with LeMessurier, and composition of the *New Yorker* piece is found in Joe Morgenstern to the author, email, January 28, 2024.

2 Forsyth, letter to the editor.

3 Mary Grace Luke to William J. LeMessurier, June 8, 1995, Papers of William LeMessurier.

4 Korman, "LeMessurier's Confession," 10.

5 Korman, "Critics Grade LeMessurier's Confession," 10.

6 Korman, "Critics Grade LeMessurier's Confession," 10.

7 Korman, "Critics Grade LeMessurier's Confession," 10.

8 Kremer, "(Re)Examining the Citicorp Case."

9 LeMessurier, "Fifty-Nine-Story Crisis."

10 Vardaro, "LeMessurier Stands Tall."

11 American Society of Civil Engineers, abstract to Morgenstern, "Fifty-Nine-Story Crisis," 23; See also Kremer, "(Re)Examining the Citicorp Case," 317.

12 "People to Know Delayed Reaction," 63.

13 Morgenstern, "Fifty-Nine-Story Crisis," 45.

14 Diane Hartley to David Billington, email, May 18, 2006.

15 BuiltWorlds, "Watch: An Epic Tale"; Ley, "Student Saves Skyscraper"; Diane Hartley to David Billington, email, May 18, 2006.

16 Berg, "Citicorp Selling," 35.

17 BXP, "Boston Properties."

18 Glanz and Lipton, "Midtown Skyscraper," A1.

19 Glanz and Lipton, "Midtown Skyscraper," A1.

20 Alberts, "Twin Towers Engineer."

21 Hickman, "Leslie E. Robertson," quoting Robertson, *Structure of Design*.

22 Weingardt, "William LeMessurier," 35.

23 Petroski, *To Engineer Is Human*, 106.

BIBLIOGRAPHY

Alberts, Hana R. "Twin Towers Engineer Blamed Himself after 9/11." *New York Post*, October 13, 2018. https://nypost.com.

American Institute of Architects. *AIA Guide to New York City*. New York: Macmillan, 1978.

American Society of Civil Engineers. Abstract to Joe Morgenstern, "The Fifty-Nine-Story Crisis." *Journal of Professional Issues in Engineering and Practice* 123, no. 1 (January 1997): 23–29.

Ammann, Othmar. "Present Status of Design of Suspension Bridges with Respect to Dynamic Wind Action." *Journal of the Boston Society of Civil Engineers* 40–41 (1953): 231–55.

Authentic History of the Lawrence Calamity Embracing a Description of the Pemberton Mill, a Detailed Account of the Catastrophe, a Chapter of Thrilling Incidents, List of Contributions to the Relief Fund, Name of the Killed and Wounded, Abstracts of Sermons on the Subject, Report of the Coroner's Inquest, &c. Boston: John J. Dyer, 1860.

"Bank Reached for Topless Neighbors." *New York Daily News*, December 14, 1976.

Banovic, S. W., T. Foecke, W. E. Luecke, J. D. McColskey, C. N. McCowan, T. A. Siewert, and F. W. Gayle. "The Role of Metallurgy in the NIST Investigation of the World Trade Center Towers Collapse." *Journal of the Minerals, Metals & Materials Society* 59 (2007): 22–30. https://doi.org/10.1007/s11837-007-0136-y.

BBC. "All Fall Down." *The Works*, April 14, 1996. Accessible on YouTube at www.youtube.com/watch?v=njeC1RmrWJo.

Berg, Eric N. "Citicorp Selling Part of Its Headquarters." *New York Times*, October 3, 1987.

Billington, David P. *The Tower and the Bridge: The New Art of Structural Engineering*. New York: Basic Books, 1983.

Boyn, Oliver, and Marie Frohling. *The Divided Berlin, 1945–1990: The Historical Guidebook*. Berlin: Christoph Links Verlag, 2012.

Brink-Johnson, Aaron, and Judy Lubin. "Structural Racism in New York City: Facts, Figures & Opportunities for Advancing Racial Equity." Center for Urban and Racial Equity, August 19, 2020.

BuiltWorlds. "Watch: An Epic Tale of How Secret Teamwork Saved Lives, in 1978." January 2, 2016. https://builtworlds.com.

Buyukozturk, Oral. "How Safe Are Our Skyscrapers? The World Trade Center Collapse." *MIT News*, September 21, 2001. https://news.mit.edu.

BXP. "Boston Properties Announces 4th Quarter and Full Year 2021 Results; Reports Q4 EPS of $1.18 and FFO Per Share of $1.55." Accessed February 10, 2024. https://investors.bxp.com.

Campbell, Robert. "Citicorp Center the Toast of New York." *Boston Sunday Globe*, May 7, 1978.

Chass, Murray. "Martin Resigns." *New York Times*, July 25, 1978.

Citi. "A Heritage of Enabling Growth and Economic Progress." Accessed October 5, 2022. www.citigroup.com.

"Citicorp Bldg. to Get 1M Wind Bracing." *New York Daily News*, August 9, 1978.

"Citicorp to Brace for Big Wind." *Boston Evening Globe*, August 9, 1978.

"Citicorp Tower Gets More Steel Bracing as Added Precaution." *Wall Street Journal*, August 9, 1978.

"Citicorp Tower Getting 'Braces.'" *New York Post*, August 9, 1978.

Clark, Gilbert B. *Hurricane Cora August 7–12, 1978*. National Hurricane Center report. Miami, FL: National Oceanic and Atmospheric Administration, 1978. www.nhc.noaa.gov.

Clinton, Senator Hillary Rodham. "Statement in Response to the National Institute of Standards and Technology's Report on the Collapse of the World Trade Center (WTC) Towers on Sept. 11, 2001, Wednesday, April 06, 2005." Skyscraper Safety Campaign, www.skyscrapersafety.org.

Corwin, Jane L., and William T. Miles. "Impact Assessment of the 1977 New York City Blackout." SCI Energy Systems, July 1978.

Council on Tall Buildings and Urban Habitat. "The Global Impact of 9/11 on Tall Buildings." Accessed August 25, 2023. www.skyscrapercenter.com.

———. Home page. Accessed August 25, 2023. www.ctbuh.org.

Crosbie, Michael J. "Hugh Stubbins, Modern Tower." *Architecture Week*, August 9, 2006.

Davis, James A. *State of Washington Department of Highways Eighteenth Biennial Report of the Director of Highways, 1938–1940*. Olympia, WA: State Printing Plant, 1940.

DeCarolis, Lee. "Citicorp Building: Who Was the Mystery Student? Commentary on William LeMessurier—The Fifty-Nine Story Crisis: A Lesson in Professional Behavior." Online Ethics Center for Engineering and Science. Accessed August 11, 2023. https://onlineethics.org.

Dempsey, Terrence, Ralph Peterson, and Jane Daggett Dillenberger. "Ralph Peterson and Jane Daggett Dillenberger." Interview by Terrence Dempsey. *MOCRA Voices*, podcast, St. Louis University Museum of Contemporary Religious Art, July 8, 2013. www.slu.edu.

Derr, Amandus. Preface to *Religion and Art in the Heart of Modern Manhattan: St. Peter's Church and the Louise Nevelson Chapel*, edited by Aaron Rosen, xix–xxi. Burlington, VT: Ashgate, 2016.

Dryansky, G. Y. "Hugh's Masterpiece." *Town and Country*, November 1979.

Duke, Lynne. "From Anger to Action." *Washington Post*, June 23, 2002. www. washingtonpost.com.

Duthinh, Dat. "Blown Away: Citicorp Center Tower Repairs Revisited." *Engineers Journal*, July 1, 2019.

Dwyer, Jim. "Didja Hear about Citicorp Tower?" *New York Daily News*, May 24, 1991.

Edes, Gordon. "Was a Newspaper Strike a Bigger Villain than Bucky Bleepin' Dent in 1978?" *Medium*, October 2, 2018. https://gewrite.medium.com.

Egan, Jack. "Citicorp Center Takes Its Place in Manhattan." *Washington Post*, October 15, 1977, E1.

"Engineering for Architecture." *Architectural Record*, mid-August 1976.

"Engineer's Afterthought Sets Welders to Work Bracing Tower." *Engineering News-Record*, August 17, 1978.

"Executives and Workmen Celebrate Topping Out of the Citicorp Center." *New York Times*, October 7, 1976.

Fitzgerald, F. Scott. *This Side of Paradise*. New York: Charles Scribner's Sons, 1920.

Flood, Joe. "Why the Bronx Burned." *New York Post*, May 16, 2010.

Forsyth, Douglass B. Letter to the editor. "In the Mail." *New Yorker*, July 10, 1995.

Friendly, Jonathan. "The Story of the Newspaper Walkout: Some Successes, Some Miscalculations." *New York Times*, November 6, 1978.

Gardiner, Stephen. "Transparently Beautiful." *The Observer*, November 23, 1980.

Gauvreau, Paul. "Current Realities and New Opportunities for Efficiency, Economy, and Elegance in Bridge Design." In *Proceedings of the IASS Annual Symposium, 2019*, 1–8. Madrid: IASS.

Glanz, James, and Eric Lipton. "A Midtown Skyscraper Quietly Adds Armor." *New York Times*, August 15, 2002.

Goldberger, Paul. "Citicorp's Center Reflects Synthesis of Architecture." *New York Times*, October 12, 1977.

———. "Diversity Marks Architecture Awards." *New York Times*, June 6, 1978.

———. "No Taint of Materialism in Church Design at Bank Center." *New York Times*, December 5, 1977.

Gottlieb, Martin. "She Draws the Interest at Citicorp." *New York Daily News*, September 19, 1977.

Hartley, Diane Lee. "Implications of a Major Urban Office Complex: The Scientific, Social, and Symbolic Meanings of Citicorp Center, New York City." BS thesis, Princeton University, April 21, 1978.

———. Presentation at the offices of Simpson Gumpertz & Heger. Washington, DC, October 14, 2019.

Harvey, Mark Sumner. "Jazz Ministry in Manhattan: The Shepherd, the Night Flock, and the First Church of Jazz." In *Religion and Art in the Heart of*

Modern Manhattan: St. Peter's Church and the Louise Nevelson Chapel, edited by Aaron Rosen, 157–82. Burlington, VT: Ashgate, 2016.

Hebert, Paul J. *Preliminary Report Tropical Storm Bess*. National Hurricane Center report. Miami, FL: National Oceanic and Atmospheric Administration, 1978. www.nhc.noaa.gov.

Hellman, Peter. "How They Assembled the Most Expensive Block in New York's History." *New York*, February 25, 1971.

Henige, Richard A., Jr. "William J. LeMessurier, 1926–2007." In *Memorial Tributes*, vol. 21, 205–9. Washington, DC: National Academies Press, 2017.

Heschel, Abraham Joshua. *The Insecurity of Freedom: Essays on Human Existence*. New York: Noonday Books, 1967.

Hickman, Matt. "Leslie E. Robertson, Structural Engineer of the World Trade Center, Passes Away at 92." *Architect's Newspaper*, February 12, 2021. www.archpaper.com.

Higgins, Michelle. "Keeping Skyscrapers from Blowing in the Wind." *New York Times*, August 7, 2015, Real Estate.

Horsley, Carter B. "Citicorp's Skyscraper 'Weathering.'" *New York Times*, March 16, 1980.

———. "A New Wrinkle on the City's Skyline." *New York Times*, September 19, 1976.

Huxtable, Ada Louise. "The New Urban Image? Look Down, Not Up." *New York Times*, January 6, 1974.

Ireland, Corydon. "A River Runs through It." *Harvard Gazette*, October 20, 2010. https://news.harvard.edu.

Jackson, Kenneth T., Lisa Keller, and Nancy Flood, eds. *The Encyclopedia of New York City*. New Haven, CT: Yale University Press, 2010.

Jacobs, Joseph. *English Fairy Tales*. London: David Nutt, 1890.

Jiang, Jian, Qijie Zhang, Liulian Li, Wei Chen, Jihong Ye, and Guo-Qiang Li. "Review on Quantitative Measures of Robustness for Building Structures against Disproportionate Collapse." *International Journal of High-Rise Buildings* 9, no. 2 (June 2020): 127–54.

Johnston, Laurie. "Church Moves in a Procession Up Park." *New York Times*, March 5, 1973.

"John White, Real Estate Agent Expert in Big Sales, Dies at 75." *New York Times*, April 2, 1995.

Kamin, Blair. "City Building: The Most Complicated Game." *CTBUH Journal* 4 (2022): 54–57.

Kleiman, Dena. "New York Cheers, and the President Signs." *New York Times*, August 9, 1978.

Korman, Richard. "Critics Grade Citicorp Confession." *Engineering News-Record* 234, no. 21 (November 20, 1995): 10.

———. "LeMessurier's Confession." *Engineering News-Record* 235, no. 18 (October 30, 1995): 10.

Kremer, Eugene. "(Re)Examining the Citicorp Case: Ethical Paragon or Chimera?" *Cross Currents* 52, no. 3 (Fall 2002): 315–27.

Landmarks Preservation Commission (New York). "Citicorp Center (Now 601 Lexington Avenue) including Saint Peter's Church." Designation List 491. December 6, 2016.

Larson, Erik. *Isaac's Storm*. New York: Vintage Books, 1999.

Lawrence, Miles B. *Preliminary Report Hurricane Ella 30 August–05 September 1978*. National Hurricane Center report. Miami, FL: National Oceanic and Atmospheric Administration, 1978. www.nhc.noaa.gov.

Lee, Peter. "The Bankers That Define the Decades: John Reed, Citibank." Euromoney, June 18, 2019. www.euromoney.com.

"Legs Centered under Each Face Carry Diagonal Braced Lower." *Engineering News-Record*, June 24, 1976.

LeMessurier, William. "The Fifty-Nine-Story Crisis: A Lesson in Professional Behavior." Lecture, National Academy of Engineering, Massachusetts Institute of Technology, November 17, 1995.

———. Papers. Frances Loeb Library, Harvard School of Design.

Lepine, Ayla. "'Upon This Rock': An Architectural History of St. Peter's." In *Religion and Art in the Heart of Modern Manhattan: St. Peter's Church and the Louise Nevelson Chapel*, edited by Aaron Rosen, 23–42. Burlington, VT: Ashgate, 2016.

Ley, Tyler. "Student Saves Skyscraper | Diane Hartley Citicorp Center Interview." December 6, 2019. www.youtube.com/watch?v=GISQfk6eN3E.

Lorenz, Edward N. *The Essence of Chaos*. London: UCL Press, 1993.

Ludman, Dianne M. *Hugh Stubbins and His Associates: The First Fifty Years*. Cambridge, MA: Stubbins Associates, in collaboration with Harvard University Graduate School of Design, 1986.

Mahler, Jonathan. *Ladies and Gentlemen, the Bronx Is Burning: 1977, Baseball, Politics, and the Battle for the Soul of a City*. New York: Picador / Farrar, Straus and Giroux, 2005.

Martinez, Jose. "Transit Strike of 1966 Remembered for Cementing Transport Workers Union's Confrontational Image." Spectrum News NY 1, January 7, 2016. https://ny1.com.

McGee, Senator Gale W. Opening statement, February 16, 1976. *Guatemala Earthquake: Hearing before the Subcommittee on Western Hemisphere Affairs and Subcommittee on Foreign Assistance of the Committee on Foreign Relations, United States Senate, Ninety-Fourth Congress, Second Session on Impact of and Relief Efforts for Guatemala Earthquake*. Washington, DC: US Government Printing Office, 1976.

Mehlman, Robert. "Elevating the Urban Environment." *Industrial Design*, May–June 1977.

Modern Steel Construction 19, nos. 1–2 (1979).

Morgenstern, Joe. "The Fifty-Nine-Story Crisis." *New Yorker*, May 29, 1995.

———. Transcript of interview with William LeMessurier. April 1993. Provided to the author by Morgenstern.

———. Transcript of interview with Leslie Robertson. March 10, 1993. Provided to the author by Morgenstern.

———. Transcript of telephone interview with Mike Reilly. March 23, 1993. Provided to the author by Morgenstern.

Morris, Deborah S. "Tips for Better, Safer Buildings, Report Recommend Updating Standards, Using Fire-Resistant Materials for Taller Structures." *New York Newsday*, June 24, 2005. Accessed at Skyscraper Safety Campaign, www.skyscrapersafety.org.

National Bureau of Standards. *Investigation of the Kansas City Hyatt Regency Walkways Collapse.* NBSIR 82-2465. Washington, DC: US Department of Commerce, February 1982.

National Institute of Standards and Technology. "FAQs—NIST WTC Towers Investigation." Created September 14, 2011, updated September 10, 2021. www.nist.gov.

National Society of Professional Engineers. "Code of Ethics for Engineers." Accessed September 28, 2023. www.nspe.org.

National Transportation Safety Board. *Highway Accident Report: Collapse of U.S. 35 Highway Bridge, Point Pleasant, West Virginia.* Report No. NTSB-HAR-71-1. December 15, 1967.

National Weather Service. "Special Request to East North Carolina and Southeast Virginia Broadcasters." National Oceanic and Atmospheric Administration, September 1, 1978. www.nhc.noaa.gov.

———. "Tropical Storm Ella Special Advisory Number 1." National Oceanic and Atmospheric Administration, August 30, 1978. www.nhc.noaa.gov.

New York City Planning Commission. *New Life for Plazas.* New York: New York City Planning Commission, 1975.

"Office Building Here to Be Largest Yet." *New York Times*, August 5, 1925.

"Office Complex to Hold Church from Which It Is Acquiring Site." *New York Times*, May 24, 1971.

Olson, Robert A., and Richard Stuart Olson. "The Guatemala Earthquake of 4 February 1976: The Social Science Observations and Research Suggestions." *Mass Emergencies* 2 (1977).

Oser, Alan S. "About Real Estate." *New York Times*, September 24, 1975.

"Outer Banks Waterspout Leaves 1 Dead." *Suffolk (VA) News-Herald*, August 1, 1978.

"People to Know: Delayed Reaction." *Modern Steel Construction*, October 2012.

Peterson, Ralph. *Life at the Intersection.* New York: St. Peter's Development Task Force, 1971.

———. *St. Peter's: A Vision for the City.* New York, 1973.

Petroski, Henry. *To Engineer Is Human: The Role of Failure in Successful Design.* New York: Vintage Books, 1992.

———. *To Forgive Design: Understanding Failure*. Cambridge, MA: Harvard University Press, 2012.

Piber, Astrid. "Megacities vs. Urban Sprawl; Densifying vs. Social Distancing." *CTBUH Journal* 4 (2021): 22–28.

"Plan for Skyscraper on Lexington Ave. Detailed by Citibank." *New York Times*, July 25, 1973.

Post, Nadine M. "World-Renowned Structural Engineer Les Robertson, ENR's Man of the Year in 1986, Dies." *Engineering News-Record*, March 1/8, 2021, 16.

Rechtien, Matthew R. "The Hyatt Regency Disaster Revisited." *Structure Magazine*, March 2014.

Red Cross. "A Brief History of the American Red Cross." Accessed November 10, 2023. www.redcross.org.

———. "Founder Clara Barton." Accessed August 10, 2024. www.redcross.org.

Reel, William. "East Side, West Side, All Over Town." *Daily News*, December 16, 1977.

Regan, Hugo. "Structural Integrity Video, 99 Percent Invisible." YouTube, January 4, 2016. www.youtube.com/watch?v=mb3LyB158YA.

Regenhard, Sally. Statement to the National Commission on Terrorist Attacks upon the United States, November 19, 2003. https://govinfo.library.unt.edu.

Rippeteau, Jane. "How Much Wind Can a Building Take? Engineers Trying to Find How Much Wind a Building Can Take." *New York Times*, April 27, 1980.

Roberts, Sam. "Roxcy Bolton, Feminist Crusader for Equality, Including in Naming Hurricanes, Dies at 90." *New York Times*, May 21, 2017.

Roberts, Siobhan. *Wind Wizard: Alan G. Davenport and the Art of Wind Engineering*. Princeton, NJ: Princeton University Press, 2013.

Robertson, Leslie E. *The Structure of Design: An Engineer's Extraordinary Life in Architecture*. New York: Monacelli, 2017.

Rosen, Aaron. Introduction to *Religion and Art in the Heart of Modern Manhattan: St. Peter's Church and the Louise Nevelson Chapel*, edited by Aaron Rosen, 1–9. Burlington, VT: Ashgate, 2016.

———, ed. *Religion and Art in the Heart of Modern Manhattan: St. Peter's Church and the Louise Nevelson Chapel*. Burlington, VT: Ashgate, 2016.

Rubin, Debra K. "NYC Builder Who Challenged Trump Dies at 92." *Engineering News-Record*, February 27, 2018, Obituary.

"Sally Regenhard, Founder of Skyscraper Safety Campaign, Recalls the Memory of Her Firefighter Son." *Yeshiva University News*, September 11, 2007. https://blogs.yu.edu.

Santosuosso, Ernie. "The Swivel-Hipped Prophet of Rock." *Boston Globe*, August 17, 1977.

Schmertz, Mildred F. "Citicorp Center as Urban Design." *Architectural Record*, June 1978.

Silverman, Steve. *Einstein's Refrigerator and Other Stories from the Flip Side of History*. Kansas City, MO: Andrews McMeel, 2001.

Stern, Robert A. M., Gregory F. Gilmartin, and Thomas Mellins. *New York 1930: Architecture and Urbanism between Two World Wars*. New York: Rizzoli, 1987.

Stetson, Damon. "Talks Intensify as Threatened Pressmen's Strike Nears." *New York Times*, August 8, 1978.

Stubbins, Hugh. *Architecture: The Design Experience*. Edited by Susan Braybrooke. New York: Wiley, 1976.

———. Papers. Frances Loeb Library, Harvard School of Design.

———. "A Skyscraper for People." *Christian Science Monitor*, November 1, 1977.

Sullivan, John. "David Billington, Scholar of Structural Art, Dies at 90." Princeton University, April 15, 2018. www.princeton.edu.

Sullivan, Patricia. "Walter B. Wriston, 85." *Washington Post*, January 21, 2005.

Tacitus. *Complete Works of Tacitus*. Translated by Alfred John Church and William Jackson Brodribb. Edited by Moses Hadas. New York: Random House, 1942.

"Transcript of Carter's Talk at City Hall." *New York Times*, August 9, 1978.

United Press International. "Citicorp to Brace for Big Wind." *Boston Globe*, August 9, 1978.

Vardaro, Michael J., Esq. "LeMessurier Stands Tall: A Case Study in Professional Ethics." AIATrust. Accessed December 3, 2023. https://theaiatrust.com.

Von Eckardt, Wolf. "Wolf Von Eckardt on Architecture: Double Manhattan." *New Republic*, January 21, 1978.

Waldorf Astoria. "History." Accessed October 31, 2023. www.waldorftowers.nyc.

Walsh, Edward, and Repps Hudson. "'Tea Dance' Hotel Disaster Toll Reaches 111." *Washington Post*, July 19, 1981. www.washingtonpost.com.

Washington State Department of Transportation. "Tacoma Narrows Bridge History—Eyewitness Accounts of November 7, 1940." Accessed September 9, 2023. https://wsdot.wa.gov.

———. "Tacoma Narrows Bridge History—Lessons from the Failure of a Great Machine." Accessed September 9, 2023. https://wsdot.wa.gov.

Weingardt, Richard G. "William LeMessurier: Builder of Elegant Cutting-Edge Structures." *Structure Magazine*, September 2012.

West Virginia Department of Transportation. "Silver Bridge." Accessed September 19, 2023. https://transportation.wv.gov.

Witcher, T. R. "From Disaster to Prevention: The Silver Bridge." *Civil Engineering Magazine*, December 2017.

Wolfe, Tom. "The 'Me' Decade and the Third Great Awakening." *New York*, August 23, 1976.

Wriston, Walter B. Papers. Tufts Archival Research Center, Tufts University.

Zweig, Phillip L. *Wriston: Walter Wriston, Citibank, and the Rise and Fall of American Financial Supremacy*. New York: Crown, 1995.

INDEX

Page numbers in *italics* indicate Figures.

103, 156, 159, 164, 172, 186–87; of
Citicorp Tower, 23; civil, 8, 39, 83, 185,
187; design of Citicorp Center, 13, 28,
29–30, 43, 46, 47–48, 61, 71; electrical, 74, 113, 152; elegance, 39; ethics,
54, 185; *Journal of Professional Issues
in Engineering Education and Practice*,
187; media with Citicorp crisis, 177,
183–89, 213n1; National Academy of
Engineering, 8, 35, 95; Online Ethics
Center for Engineering and Science, 54;
at Princeton University, 39; safety with
redundancy, 70–71, 85, 90, 92–93, 94;
structural, 6, 7, 16, 17, 52, 54, 58, 79,
83, 85, 89, 106, 184; team for WTC, 15,
101–2, 111, 145, 190; *To Engineer Is
Human*, 191; women in, 39
Engineering News-Record (magazine), 49,
50, 102, 141–42, 183, 185, 198n41
engineers: audit of Citicorp Tower by
design review, 155–60, 167–68, 172–
74, 176, 178, 211n5 (chap. 17); 212n2
(chap. 19); LeMessurier, William, as,
7, 16–17, 23; Missouri Board for
Architects, Professional Engineers and
Land Surveyors, 91; National Society of
Professional Engineers, 92; New York
Society of Professional Engineers, 185
Ernst & Ernst, 51
error in design, Citicorp Tower: audit
by review engineers, 155–60, 167–68,
172–74, 176, 178, 211n5 (chap. 17),
212n2 (chap. 19); braces connected by
bolts instead of welds, 35–36, 55, 57–62,
69–71, 75–77, 103, 105, 107–8, 111,
116, 134, 148, 151–52, 155, 157, 173,
177, 181, 185–86, 189, 202n9 (chap. 6);
braces remediated with welds, 105, 111,
115–17, 120, 126, 137, 141–42, 145–49,
151, 153–54, 156–61, 163–64, 167;
corrective plan, 105–9, 111–13, 115–21,
135–36, 142, 145–48, 171, 175–76,
210n19, 211n2 (chap. 16); evacuation
plans, 105, 119, 120, 121, 125, 126,
130–31, 139, 149–51, 153–54, 163, 168,
186; flooring, 157–61, 212n2 (chap. 19);
insurance companies, 99, 128, 168, 170;

lawyers, 100–106, 120–21, 126, 128,
168–71, 176–80; LeMessurier, William,
with accountability, 69, 71, 73–80, 94,
96–98, 103, 129, 135–36, 170, 177,
181, 185–87, 205n9 (chap. 8); public
statement on "structural improvements,"
133–39, 141–42, 143–44, 146, 176–77,
186; quartering wind and structural
vulnerability of, 98–99, 103–4, 107–8,
111, 133; settlement claim, 179–81;
trench header issue, 158–61, 165; white
lies about, 135, 136–37, 139, 141–42,
177–78, 186. *See also* bolts; braces;
"Project SERENE"
The Essence of Chaos (Lorenz), 212n2
(chap. 18)
ethics: code of, 92, 186; engineering classes,
185; Online Ethics Center for Engineering and Science, 54; professional, 96,
97, 181, 182, 185, 187; white lies, 135,
136–37, 139, 141–42, 177–78, 186
evacuation plans: American Red Cross,
126, 130–31, 139, 150–51; Citicorp
Tower, 105, 119, 120, 121, 125, 126,
130–31, 139, 149–51, 153–54, 163,
168, 186; media questions about, 139;
with secret canvassing of perimeter,
150–51; Skyscraper Safety Campaign
with, 65

face wind. *See* perpendicular wind
failure: electrical grid, 121–22, 208n1; of
oversight, 46, 67, 71, 79, 186, 187
failure, structural: amphitheater at Fidena,
80–81; analysis, 79, 85; bridges, 44,
81–86, 91, 92; bridges and hypothetical
collapse, 183, 184; building collapse,
64–67, 70–71, 77–81, 87–92, 94, 125,
130–31, 135, 173, 177, 181, 190,
203n12; causes and reasons for, 91–93;
Citicorp Tower with probability of, 96,
130–31, 146; deaths, 64–65, 80–81, 84,
87, 88, 90, 92, 129; eyewitness accounts,
80–81, 82, 84, 87–88, 89; South Fork
Dam, 129; Twin Towers, 65–66, 181,
190, 203n12
Farquharson, F. B. "Bert," 81–82

Medical Chambers building, xvi, xix, 13, 14
Mehlman, Robert, 34
meteorologists, 102, 109, 118, 137, 140, 162
Michel, John C., xv
misconduct, 91, 92
Missouri Board for Architects, Professional Engineers and Land Surveyors, 91
Missouri Court of Appeals, 91, 94
MIT. *See* Massachusetts Institute of Technology
Morgenstern, Joe, 181, 209n19 (chap. 14); with Citicorp engineering crisis, 183–86, 187, 188, 189; on Citicorp Tower, 23; with *New Yorker*, 184–87, 188, 189, 213n1; with Robertson, Leslie, on Citicorp Center, 206n16
Morrison, Peter, 169, 180
MTS Systems Corp., 28, 113, 116, 118, 126–27, 137
Muller, J. Henry, xxi–xxiii, 3–6, 8, 12–14
Murdoch, Rupert, 145

National Academy of Engineering, 8, 35, 95
National Association of Architectural Metal Manufactures Standard WL-10-67, 201n18
National Bridge Inspection Standards Program, 86
National Bureau of Standards, 90
National Construction Safety Team Act, 66
National Hurricane Center, 140
National Institute of Standards and Technology (NIST), 65–67, 69–70, 203n13
National Science Foundation, 25
National Society of Professional Engineers, 92
National Transportation Safety Board (NTSB), 85
National Weather Corporation, 117, 127, 137
National Weather Service, 117, 118, 164, 165
negligence, 87, 91, 92, 168
New Jersey Institute of Technology, 54, 189

newspaper strike (1978), New York City, 144–45, 154, 183
New York and New Jersey Port Authority, 11
New York Central Railroad, xv
New York City: architecture, 11; culture, 4, 5, 6, 11, 12, 35; economy, xviii, 10, 32, 34, 122, 132–33; newspaper strike of 1978, 144–45, 154, 183; power blackout, 122, 208n1; transit strike of 1966, 122–23, 208n1; zoning, 4–5, 14, 19, 22, 23, 40
New York City Building Code, 29, 44, 68, 79, 135–36, 177–78, 202n10 (chap. 6)
New York City Department of Buildings, 66, 135, 136, 137, 178
New York City Fire Department, 63–64
New York City Loan Guarantee Act (1978), 132–33
New York City Police Department (NYPD), 123–24, 151, 208n1
New York Daily News (newspaper), 34, 46, 51, 138, 139, 141, 145
New York Department of Highways, 142
New Yorker (magazine), 184–87, 188, 189, 213n1
New York Post (newspaper), 141, 145, 197n2
New York Pressmen's Union, 144–45, 154, 183
New York Society of Professional Engineers, 185
New York Stock Exchange, 207n23
New York Times (newspaper), 60, 132, 143, 183; Citicorp Center in, 24, 34; Citicorp in, 23; Citicorp Tower in, 20, 189; with newspaper strike of 1978, 144, 145; St. Peter's Lutheran Church in, xxiii; on wind disturbances in high-rise buildings, 26–27
New York Yankees, 60, 145, 183
9/11 (September 11, 2001), 63–67, 70–72, 189–90
9/11 Independent Commission, 65–66
NIST. *See* National Institute of Standards and Technology
Nixon, Richard, 20, 110

ABOUT THE AUTHOR

MICHAEL M. GREENBURG is a practicing attorney and the author of *Peaches and Daddy: A Story of the Roaring 20s, the Birth of Tabloid Media, and the Courtship That Captured the Heart and Imagination of the American Public*; *The Mad Bomber of New York: The Extraordinary True Story of the Manhunt That Paralyzed a City*; *The Court-Martial of Paul Revere: A Son of Liberty and America's Forgotten Military Disaster*; and *This Noble Woman: Myrtilla Miner and Her Fight to Establish a School for African American Girls in the Slaveholding South*.

www.ingramcontent.com/pod-product-compliance
Lightning Source LLC
Chambersburg PA
CBHW021024190925
32857CB00012B/29/J